Vienna Circle Institute Yearbook

Institute Vienna Circle, University of Vienna
Vienna Circle Society, Society for the Advancement of Scientific World Conceptions

Volume 20

Editorial Address
Wiener Kreis Gesellschaft
Universitätscampus, Hof 1
Spitalgasse 2-4, A–1090 Wien, Austria
Tel.: +431/4277 46501 (international) or 01/4277 46501 (national)
Email: ivc@univie.ac.at
Homepage: http://univie.ac.at/ivc/

More information about this series at http://www.springer.com/series/6669

Friedrich Stadler
Editor

Integrated History and Philosophy of Science

Problems, Perspectives, and Case Studies

Editor
Friedrich Stadler
Institute Vienna Circle
University of Vienna
Vienna, Austria

ISSN 0929-6328 ISSN 2215-1818 (electronic)
Vienna Circle Institute Yearbook
ISBN 978-3-319-85106-8 ISBN 978-3-319-53258-5 (eBook)
DOI 10.1007/978-3-319-53258-5

Printed on acid-free paper

This Springer imprint is published by Springer Nature
The registered company is Springer International Publishing AG
The registered company address is: Gewerbestrasse 11, 6330 Cham, Switzerland

Contents

Editorial

This volume contains a selection of the papers presented at the Fifth Conference on "Integrated History and Philosophy of Science" (&HPS) organized by the Institute Vienna Circle at the University of Vienna, June 26–28, 2014.[1]

The organizers were pleased to host just this conference, given the related research focus and scholarly environment at the University of Vienna, e.g., the running master's program HPS and the doctoral program "The Sciences in Historical, Philosophical, and Cultural Contexts."[2] In addition, the Institute Vienna Circle had already co-organized a conference on "Wissenschaftsgeschichte und Wissenschaftsphilosophie" (History of Science and Philosophy of Science) at the University of Vienna in 2011, whose proceedings were published in the journal *Berichte zur Wissenschaftsgeschichte*.[3]

From my personal point of view, it was most reasonable to have one of these &HPS conferences in Vienna, given the denomination of my own chair: History and Philosophy of Science.[4] In the meantime we experienced a big conference on Ernst Mach (1838–1916) on the occasion of the centenary of his death, which – once more – was a variation of the interdisciplinary research field history and philosophy of science, with special contributions to Mach's significance for an integrated HPS.[5]

My thanks go to the &HPS Committee members (chaired by Don Howard and John Norton) for accepting our proposal and for acting as the Program Committee (selecting 33 speakers out of more than 150 submissions, which indicates the high

[1] &HPS website: http://www.pitt.edu/~pittcntr/About/international_partnerships/andHPSpage.html

[2] http://dk-sciences-contexts.univie.ac.at/

[3] *Berichte zur Wissenschaftsgeschichte*. Organ der Gesellschaft für Wissenschaftsgeschichte. Issues 2/2012, 3/2012, 1/2013.

[4] Friedrich Stadler, "History and Philosophy of Science," in: *Berichte zur Wissenschaftsgeschichte* 3/2012, 217–238. The abridged English version "History and Philosophy of Science – Between Description and Construction" in: *New Directions in the Philosophy of Science*. Ed. By Maria Carla Galavotti, Dennis Dieks, Wenceslao J. Gonzalez, Stephan Hartmann, Thomas Uebel, Marcel Weber. Cham-Heidelberg-New York-Dordrecht-London: Springer 2014, 747–767. (The Philosophy of Science in a European Perspective, Vol. 5).

[5] http://mach16.univie.ac.at

competition and pleasing interest in HPS in general), the cosponsoring Faculty of Philosophy and Education (represented by its dean Elisabeth Nemeth and my colleague Martin Kusch), and the staff members of the Institute Vienna Circle, esp. Sabine Koch, Karoly Kokai, and Robert Kaller.

Institute Vienna Circle
University of Vienna
Vienna, Austria
Friedrich.Stadler@univie.ac.at

Friedrich Stadler

Part I
Integrated History and Philosophy of Science. Contributions from the 5th Conference

Chapter 1
Metaphysics and the Unity of Science: Two Hundred Years of Controversy

Richard Creath

Abstract Carnap's rejection of metaphysics and his embrace of the unity of science are closely intertwined. Carnap is clear about his specific target in metaphysics and about why he rejects it. Surprisingly, on his mature position he does not show us that we cannot be realists, or nominalists, or idealists, etc., but rather how we can. Carnap directs his remarks on the unity of science toward a specific family of claims, prominent in the early twentieth century, namely that the natural sciences are to be sharply divided from the human sciences. Windelband wrote a famous and influential paper that defends such a division. A close look at this paper shows how Carnap's position presents the two-kind-of-science view with a dilemma: Either the attempt to divide the sciences in that particular way fails, or the division crosses the boundary into metaphysics.

Rudolf Carnap's rejection of metaphysics and his embrace of the unity of science are intimately intertwined. They may not seem that way at present, but that is because both are now regularly misunderstood. In this paper I want to bring both parts of Carnap's view into sharper focus, that is, to see what Carnap actually had in mind and why he took the stands he did. In the first part of the paper I will review Carnap's attitudes toward metaphysics. I will review a common misunderstanding of Carnap's views on metaphysics and then show that Carnap is clear and specific about what he means by 'metaphysics', whom he sees as guilty, and why he rejects their sort of work. And I will show that by the mid 1930s Carnap came to a surprising response to the enterprise he rejected.

In the second part of this paper I will look at Carnap's ideas on the unity of science. After reviewing several common misunderstandings I will show how Carnap's discussion thereof is also directed at a definite target, one that insists that there are two sorts of science that differ is specific ways. While there are many writers who advanced this two-sorts-of-science view (I will call it the dyadic view for short), I will concentrate on the views of Wilhelm Windelband. This issue of the unity of science deserves more attention because the view that Carnap rejects, while little known to American philosophers, is subtle and formidable. It is also part of a very

R. Creath (✉)
Arizona State University, Tempe, AZ, USA
e-mail: Creath@asu.edu

F. Stadler (ed.), *Integrated History and Philosophy of Science*, Vienna Circle Institute Yearbook 20, DOI 10.1007/978-3-319-53258-5_1

long tradition that dates from at least the beginning of the nineteenth century and continues on even today. In Carnap's response, we will see the connection to the issue of metaphysics.

1.1 Metaphysics

Carnap's rejection of metaphysics, especially in the early 1930s, is, to say the least, vigorous. He wants us to understand that he is not calling it doubtful, or false, or even logically false. He is not saying that it is of no earthly value for any practical purpose, though presumably this would follow from what he does say of it. Instead, he says that metaphysics is utterly meaningless. It is without cognitive content of any kind; it is gibberish. Unsurprisingly, many contemporary philosophers who count themselves as metaphysicians feel slighted by this. They take him to be rejecting the entire field and everything in it as gibberish. In fact, Carnap is not rejecting the whole field – quite. And the way in which he is not rejecting the whole field may be surprising. He might still reject many specific things that metaphysicians have written. But then what metaphysician doesn't?

If it is a mistake to think that Carnap's criticism must apply to the whole field, it is also a mistake to think that it applies only to work of long ago. It is true that Carnap's most notorious paper against metaphysics, "Overcoming Metaphysics Through the Logical Analysis of Language" (Carnap 1932a/1959), prominently featured Heidegger as an example of the sins of metaphysics. So one might mistakenly reason: OK, perhaps some of Heidegger's claims like "Nothing noths." Or "Anxiety reveals the nothing." are really meaningless. But what does that have to do with us? We are completely unlike Heidegger. We are trained in logic, and even our metaphysics is highly technical. We would never make those mistakes. It is true that contemporary analytic metaphysicians are unlike Heidegger in several respects that Carnap takes to be important. But it would be premature to think oneself safe from whatever criticism Carnap might have of metaphysics simply on the grounds of dissimilarity to Heidegger. We need to articulate what those criticisms were, and to that we now turn.

For Carnap, metaphysics is an enterprise based on the idea that by philosophic or other non-empirical means one can know truths about the world that are prior to or lie behind or are deeper than the truths to which empirical science can aspire. The idea that there are such deeper truths that are inaccessible to science perhaps goes back to Kant's discussion of things-in-themselves. Kant said that science could never know these things-in-themselves (except that there were such), and so we were to refrain from speculating about them. Well, German Idealists rush in where Kantians fear to tread. There aren't things-in-themselves exactly (It was the "in-themselves" part that they objected to.), but there was a domain of deeper truths about which we could have genuine knowledge. And the road to this knowledge runs through philosophy, i.e., metaphysics, rather than through empirical science.

Besides the German Idealists, Fichte, Schelling, and Hegel, Carnap also names Bergson and, notoriously, Heidegger under the heading of "metaphysicians". Actually, though we need not go into this here, he has much more extensive lists of metaphysical philosophers, even indicating their individual degrees of guilt. In his "Intellectual Autobiography" he adds one more name to the list, his sometime colleague, Mortimer J. Adler. Carnap tells the story of a lecture that Adler gave in which

> [h]e declared that he could demonstrate on the basis of purely metaphysical principles the impossibility of man's descent from "brute", i.e., subhuman forms of animals. I had, of course, no objection to someone's challenging a widely held scientific theory. What I found startling was rather the kind of arguments used. They were claimed to provide with complete certainty an answer to the question of the validity or invalidity of a biological theory, without making this answer dependent upon those observable facts in biology and paleontology, which are regarded by scientists as relevant and decisive for the theory in question. (Carnap 1963, 42)

Carnap uses this as an example of a kind of "cultural lag" that he found at the University of Chicago. But in its confidence in philosophic methods and its dismissal of empirical ones, it is hard not to see Adler as the perfect example of what Carnap means by 'metaphysics'.

And what Carnap meant is largely what these metaphysicians themselves meant. The writers that Carnap was talking about defined their own approach in opposition to empirical science, both in terms of its methods and in terms of its results: Metaphysical methods are not at all empirical, and the resultant knowledge is deeper than and concerns a reality that lies behind the world of empirical science. Some metaphysicians, like Heidegger, devalue not only empirical science but logic as well. But see also (Naess 1965/1968).

Against this Carnap argues that with a metaphysical approach each writer builds his or her own system among which there are "wearisome controversies" that never get anywhere. One system may become temporarily fashionable, but it doesn't last. There is nothing that would count as a significant test between such metaphysical systems as Realism, Materialism, Idealism and so on. Science by contrast, is cooperative and progressive. So science gets somewhere, and metaphysics doesn't. This, by the way, is a thoroughly historical argument.

In the early 1930s Carnap saw empirical science as a neutral core to which all metaphysicians could agree and the various metaphysical positions as various additions to that. Carnap illustrates this by giving a parable in *Pseudoproblems* (Carnap 1928b/1967, 333) about the Realist and the Idealist geographers who go off on an expedition to find out whether a certain mountain exists. They come back with the same report about the mountain's height, shape, and location. This is hardly surprising. They used the same instruments for determining latitude, longitude, and elevation. The two geographers agree on the report, but the Realist adds that the mountain that they have located and measured is also real. The Idealist, by contrast, insists that the mountain itself is not real, and only our perceptions are real. No evidence could settle this dispute. All the evidence they have is summarized by the neutral core. Where they disagree is over what to infer from that evidence, and this is not an empirical question. If this is the metaphysicians' own conception of their enterprise, then (to remodel a remark by Quine): Why all this metaphysical make believe? Why not settle for empirical science?

There was to be a subtle but important change in his attitude toward metaphysics as he moved to his mature philosophy in the mid-1930s. In the new regime Carnap recommends that we reconstrue erstwhile metaphysical positions, such as Realism and Idealism, as proposals for structuring the language of science. Philosophic work would thus comprise the engineering tasks of explicating our current linguistic framework, devising new ones, and exploring their virtues and vices for various specific practical purposes. The surprise is that Carnap thus shows us how to be a Realist or Idealist or Platonist or nominalist, that is, how to be a metaphysician, in a sense that he finds unobjectionable. This is what I meant earlier that Carnap did not dismiss the field of metaphysics entirely, just one dominant conception of it. Just as there can be legitimate analogs of the old metaphysical positions, it is, alas, possible to make mistakes that are analogs of the old ones: One might think that there is one uniquely correct linguistic framework, and one might think that we have some power of metaphysical insight that reveals which framework is correct.

So what then would be Carnap's assessment of contemporary analytic metaphysics? The answer undoubtedly would be mixed. Contemporary analytic metaphysics generally lacks the dismissive attitude toward empirical science that characterized much of nineteenth and some early twentieth century metaphysics. And happily much current metaphysics is highly logic based, and it is often uneasy with claims about deep insight. But too often this unease is soothed by disguising the claims to insight by hiding them under a different name, like 'intuition' that covers both problematic uses and completely innocuous ones. Giving metaphysical insight a different name does not change anything. So some of the old errors can get smuggled in with the result of controversies that never get anywhere. But Carnap's point is that these errors don't have to be repeated.

Carnap's mature advice for a remedy can be summarized in a single word: clarity. State clearly whatever you propose in the way of rules for inference, for observation, and for anything else. State them in full explicit detail so that we can get them out in the open so we can see how they work.[1] The historical record of the approach Carnap calls metaphysical is not good. The historical record of judgments made directly on the basis of publically accessible observations is much better. There is wide and durable agreement about the observational judgments themselves, and there is an emerging theory explaining the connections between the events observed and the observational judgments. These judgments also form the basis of broad and useful theories. And we can specify the respects in which they have improved. This, of course, gives no absolute guarantee. But it does give pragmatic grounds for using those observations for now, for pushing the theorizing and testing further, and for using the best of our empirical theories as a basis for practical action. I don't think we can, or should, ask for more.

But what does all this have to do with the unity of science? A lot, as it happens. To this we now turn.

[1] This, of course is the Principle of Tolerance, first stated as such in (Carnap 1934/1937, 51).

1.2 The Unity of Science

First, I will review what our contemporaries have imagined Carnap was saying, and then I will turn to the tradition Carnap rejects and to Carnap's response.

If many contemporary metaphysicians are annoyed by Carnap's apparent rejection of their whole field, his views on the unity of science provoke less annoyance than indifference from today's philosophers including philosophers of science. There was a brief flurry of activity a few years ago extolling the "disunity of science", but little of that had to do with Carnap (Gallison and Stump 1996 and also Creath 1996) The result of this indifference is that contemporary writers have often just attributed to Carnap whatever they imagine he might have meant.

Sometimes it is thought that unity requires that all the truths of all the sciences be logically derivable from those of physics. Carnap does consider such a possibility but says that it is not part of the unity he is claiming for science. That is a matter that should be decided by the evidence that is not yet in. Sometimes it is suggested that Carnap is claiming that science is unified ontologically, that all the objects of science are all physical or that they are all phenomenal. Such ontological interpretations undoubtedly stem from Quine's influential writings (Quine 1951, 1971), but more recent scholarship shows that ontological parsimony is far from Carnap's central concerns (Friedman 1987; Richardson 1998) Alternatively, since Carnap did talk about the unity of the language of science, some have thought that he could be refuted by showing that the various sciences have different technical vocabularies. (Suppes 1978) But this is just a serious misunderstanding of what Carnap means by 'unity of the language'. Finally, some would interpret the unity of science methodologically. Some such interpretation is correct, but it is also easy to seize on the wrong methodological features. It is easy to think that the day-to-day practices in all sciences must copy those in physics or, worse, copy some one method that is thought to be *the* method of physics. Sometime the imagined unity of method is silly: If physics is a lab science then so must be sociology. If physics does not use public opinion surveys then sociology must not either. Such is the picture of the unity of science foisted on Carnap by others. These various interpretations: logical, ontological, linguistic, and methodological, all miss the mark or are prone to do so. But the best way to avoid being misled is to see what is really going on.

1.3 The Dyadic Tradition

In embracing the unity of science, Carnap and his friends are rejecting a specific family of views, widespread in the early twentieth century, about how the sciences are divided. This set of views, represented prominently by Wilhelm Windelband,

Heinrich Rickert and by Wilhelm Dilthey[2] among others, holds that there is a deep division between the Naturwissenschaften and the Geisteswissenschaften. The former are the natural sciences such as physics and chemistry, and I will turn in a moment to the question of what sciences the Geisteswissenschaften would encompass. The basic rationale for the division is that the natural sciences seek laws and have no interest in individual cases. The Geisteswissenschaften, by contrast, deal with particular historical objects, events, and processes. This plays out differently for the various authors, but for some this concentration on historical individuals is supposed to make room for a treatment of values and human freedom, much as Kant's things-in-themselves had done.

Note that the view to which Carnap addresses his work on the unity of sciences is not that there are many different kinds of science, but that there are two. And as we shall see, these two are distinguished from each other in a particular sort of way. The general idea that there are two kinds of science, the Naturwissenschaften and the Geisteswissenschaften, we shall call the dyadic tradition.

At the time, the fields that were thought to best exemplify the Geisteswissenschaften were most often history, philosophy, and philology. The word 'philology' almost never appears in American university catalogs, so as a field of study philology is almost completely unknown there. It is just the study of languages, texts, and the cultures that go with them, especially the study of ancient languages, texts, and cultures. Now in the US the domain that includes as paradigmatic examples history, philosophy, and the study of the text, especially ancient texts, is called the humanities, not the social sciences as 'Geisteswissenschaften' is sometimes rendered. Even 'humanities' is not quite right as a translation for the Geisteswissenschaften might also include humanistically understood politics, cultural anthropology, and social theory. And there is no exact boundary that all those who defended the distinction would agree on. So perhaps the best translation for 'Geisteswissenschaften' would be 'human sciences'. Admittedly, an English speaker would generally not use the word 'sciences' in reference to the humanities but could admit that the humanities could be considered an organized body of knowledge and hence as science in a broader sense. In any case, 'human sciences' is the phrase that we will use from here on. No doubt it is translational difficulties such as these that lead English speaking philosophers of science to assume that the unity of science is about uniting physics with biology, psychology, and sociology rather than uniting the natural sciences generally with the human sciences including the humanities.

[2]That Carnap rejected some of the ideas of these three does not preclude his having been influenced by them on other matters. Thomas Mormann has argued that Rickert influenced especially Carnap's early philosophy (Mormann 2006 and see also Gabriel 2007) And Christian Damböck has been showing the influence of Dilthey (Damböck 2012).

1.4 A Defense of the Dyadic Tradition

Rather than immediately turning to how Carnap opposes the idea that there are these two radically different kinds of knowledge, that is two kinds of science, I want to suggest that the dyadic tradition that Carnap rejects is both subtle and formidable, especially in the early twentieth century (See Anderson 2012). It should not be underestimated. It is formidable in part because German scholars in the humanities had been active in the nineteenth and early twentieth centuries and pointed with pride to their achievements. And they had achieved much especially in classics. There were first rate philological works, notably critical editions of classical texts, commentaries, organized collections of fragments of later Hellenistic work, especially that of the Stoics. Carnap was fully aware of this work. His uncle was Wilhelm Dörpfeld, a distinguished archeologist, who was brought in to clean up the mess of Schliemann's excavations of Troy and to give academic respectability to the project. As a child Carnap himself had gone with his uncle on digs in Greece.

Beyond the philological work there were rich and detailed histories (Cf. Windelband 1893/1901/1958 and Cassirer 1910/1923). And the flood of highly regarded philosophic work was notable for both the sheer size and the evident systematicity of the individual contributions. Some of this work is undeniably outstanding. On the surface none of this work looks like a laboratory science. None of it is particularly mathematical. And laws of history, philosophy and philology seem to be in short supply.

1.5 Windelband

The general rationale for contrasting the natural and the human sciences becomes clearer when we look at specific attempts to argue for the division. The obvious place to start is with Windelband's "History and Natural Science" (Windelband 1894/1980). It is relatively clear and succinct.[3] It was enormously influential on the discussion that followed. And we also know that Carnap read this paper. In order to treat Windelband in some depth, I will make only brief reference to other writers such as Rickert and Dilthey.

Windelband begins by distinguishing the non-empirical sciences of mathematics and philosophy from the empirical ones, and it is within the latter that he draws the natural/human sciences distinction.[4] In discussing Windelband's attempt to distinguish two kinds of sciences, it will be useful for us to do so under four headings: (1) laws vs unique particulars, (2) abstraction vs perceptuality, (3) values, and (4) human indeterminacy and freedom.

---

[3] This is not the case for Rickert. See (Rickert 1896-1902/1986 and 1899/1962).

[4] In this Windelband obviously differs from the discussion above in which philosophy was taken as a paradigm case of the human sciences. And not everyone in his tradition followed Windelband in this.

1.5.1 Windelband: Laws vs. Unique Particulars

The aim of the natural sciences, Windelband says, is to discover laws. By contrast most of the disciplines called human sciences are concerned with a temporally bounded domain, that is, with concrete, historical particulars. As a consequence, the natural/human science divide can be thought of as the natural/historical science divide, and psychology falls unambiguously within the natural sciences.[5] Windelband also says that while we have a logic that deals well with the natural sciences, we lack such a logic that would apply to particular individuals.[6]

This would be a sharp contrast between the natural and human sciences, but Windelband no sooner stated the distinction than he proceeds to qualify it, indeed to undermine it. In a passage that would have been right at home in Hempel's "The Function of General Laws in History" (Hempel 1942), Windelband points out that any explanation of a historical event must invoke laws as well. Laws are not enough, of course. Statements of what we would call initial conditions must supplement those laws. We are left then with two elements of the human or historical sciences. Even this would be enough to distinguish them from the natural sciences if it were true that the natural science ignored initial conditions entirely. But it isn't.

Just as both laws and initial conditions are needed in historical explanation, they are also both needed for explanation within physics. Moreover, statements of particular events are needed in natural science for the confirmation and testing of theories and as predictions as well. Every experiment is a particular event. Every lab book is a historical record of particular events in the laboratory. And every claim that one's science is making progress is a historical comparison of particular theories developed at particular times. Given all this, we have two elements, laws and initial conditions, within every science and not a division at all between two kinds of science (See also Creath 2010).

1.5.2 Windelband: Abstraction vs. Perceptuality

Windelband's further claim is that the natural sciences are biased in favor of abstraction to the point that perceptual qualities are omitted entirely. Windelband goes further than saying that the natural sciences are often abstract or that they are biased in favor of abstraction. He says that the natural sciences are wholly uninterested in individual events including observations:

> But the more we strive for knowledge of the concept and the law, the more we are obliged to pass over, forget, and abandon the singular fact as such. (181)

[5] For Dilthey, psychology falls within the human sciences.

[6] Carnap responds by noting in the *Aufbau* (Carnap 1928a/1967, 23–24) that the logic of relations developed by Whitehead and Russell is ideally suited to satisfy such a need. Cassirer makes a similar point against Rickert (Cassirer 1929/1957, 348).

And again:

> Consider the single perceptual datum which appears and disappears. In genuine Platonic fashion, the natural sciences ignore this datum as a negligible and insubstantial appearance… . From the colorful world of the senses, the natural sciences construct a system of abstract concepts[,] … a silent and colorless world of atoms in which the earthly aura of perceptual qualities has disappeared completely: the triumph of thought over perception. Utterly indifferent to the past, the natural sciences drop anchor in the sea of being that is eternally the same. They are not concerned with change as such, but rather with the invariable form of change. (179)

If the bias of the natural sciences is toward abstraction, the bias of the human sciences is in favor of perceptuality (Anschaulichkeit), though this perceptuality is not limited to what one can learn from the anatomical eye. Windelband suggests that such limited perception would be passive. He says that there is also perception via "the eye of the mind" (179) and speaks in the same breath of "the spontaneous faculty of perception" and "perceptual fantasy" (179) as though this were a form of perception open to the human sciences but not to the natural sciences. Thus a special form of perception may be involved.

What would Carnap make of all this? Well, he can hardly be accused of neglecting observation in his account of physics, psychology, and other sciences. And he would, of course, resist the idea that the natural sciences lack any concern for perceptual properties. Returning to Windelband, if his claim about mind's-eye perception is only that imagination is needed in the framing of hypotheses, Carnap would be happy to agree but insist that this holds for both the natural and human sciences. If the idea of mind's-eye perception is that such perception is a source of validation for scientific hypotheses, a source that is available in the human sciences but not the natural sciences, then Carnap would resist. He would, and did, resist as well any idea that there is a form of perception open to the human sciences that is denied to the natural sciences. In fact that is precisely the argument he makes both in the *Aufbau* (Carnap 1928a/1967, 23–24, 89–92, and 230–233) and in his later unity of science works. (especially Carnap 1938, but also 1932b/1959 and 1932c/1959) For Carnap the unity of the language of science is just that all scientific concepts are to be unpacked in terms of the concepts involved in publically accessible observation. In the *Aufbau* this appears as his claim that the evidence base for the human sciences consists entirely of claims about either physical or psychological objects, events, and processes. (Carnap 1928a/1967, 89–92 and 230–233).

1.5.3 Windelband: Values

For Windelband, there is a big payoff for the idea that the human sciences deal with particular, even unique, objects. For only in the utterly unique is there any value. And Windelband illustrates this claim by saying that the line "She is not the first." is one of the "most terrifying" in *Faust* (182), that there is "horror and mystery" in the idea of the *Doppelgänger* (182), and that the idea of eternal recurrence is "painful" and "dreadful". (182) The reason given for this last is that "Life is debased when it has already transpired in exactly the same way numerous times in the past and will be repeated again on numerous occasions in the future." (182) Windelband's moral stance is clear, but he does not say what basis he has for his moral judgments. Nor does one spring to mind. Nonetheless, Windelband plainly suggests that the human sciences get us to a domain of moral knowledge beyond the reach of the evidence available in the natural sciences.

This idea that the uniqueness of objects is somehow linked to the possibility of value is a suggestion that would be picked up by other writers in the dyadic tradition, such as Rickert. But the argument for the position is not clear enough that it really calls for a response. Carnap's views on value were complicated enough in the 1930s and sufficiently separated from his unity of science concerns that it would not be useful for us to address them here.

1.5.4 Windelband: Human Indeterminacy and Freedom

Finally, Windelband's discussion of laws and initial conditions led him to a conclusion even more striking than a connection between uniqueness and value. Recall that he concedes that laws are essential parts of historical explanations. But, he adds correctly, the statements of initial conditions do not follow logically from the laws alone. Immediately after making this point he says

> General laws do not establish an ultimate state from which the specific conditions of the causal chain could ultimately be derived. It follows that all subsumption under general laws is useless in the analysis of the ultimate causes or grounds of the single, temporally given phenomenon. (184)

Conceivably, Windelband is denying that we can get to first causes from laws alone. More likely, by saying that general laws are useless he means only to reformulate his claim that the laws by themselves to not entail any statement of individual events. So understood, the claim is reasonable. But Windelband immediately draws a conclusion from this that is, at the very least, puzzling:

> Therefore, in all the data of historical and individual experience a residuum of incomprehensible, brute fact remains, an inexpressible and indefinable phenomenon. Thus the ultimate and most profound nature of personality resists analysis in terms of general categories. From the perspective of our consciousness, this incomprehensible character of the personality emerges as the sense of the indeterminacy of our nature – in other words, individual freedom. (184)

This is not the conclusion of a long and subtle argument. It is the argument in its entirety. It is also a non sequitur.

In any case the claims of the argument seem to be very substantial. If it were a result within the human sciences, then those sciences would indeed be special in requiring evidence of an extraordinary sort. If, however, Windelband intends it as the conclusion of a philosophic argument (and the text suggests that this may be the case), then as we shall see later, the argument is invalid and does nothing to establish the distinctness of the human and natural sciences.

Contrasting the natural sciences with some other sort of knowledge long predated Windelband, Rickert, and Dilthey (See Berlin 1996, 1998, and 2000 and Lilla 1993). The German Idealists that Carnap explicitly picked out as metaphysicians also distinguished the Naturwissenschaften and the Geistenwissenschaften, though their distinction is not precisely the same as in Windelband et al. A less science friendly analog appears in Bergson and Heidegger. And a version of these distinctions and arguments even shapes debates today within at least American universities between the humanities and the natural sciences. Since the general idea has been around for two hundred years, it would be optimistic to think that we can quickly put it aside or resolve the controversy. But we can at least step back to see Carnap's response as a whole and to see how it is intertwined with the issues of metaphysics.

1.6 A Dilemma

We saw that Windelband was happy to agree that history and the other human sciences are in fact empirical. Others in the dyadic tradition, such as Dilthey, did the same. And Dilthey is even explicit that metaphysics is to be avoided. But we also saw that Windelband suggests that beyond ordinary observation there may be a special sort of perception that applies only in the human sciences.[7] The point is important, for on it hinges the success of the dyadic tradition.

Carnap of course would say that the dyadic tradition is mischaracterizing the natural sciences. These sciences are not completely uninterested in individual objects, events, and processes and they do not abstract entirely away from perceptible features of things. But the issue of whether there is a special kind of perception involved in the human sciences is more important. Insofar as it restricts itself to ordinary observation, the dyadic tradition does not succeed in showing that the natural and human sciences are relevantly different. If, however, it accepts a special sort

[7] Dilthey makes a similar suggestion. Perhaps it is more than a suggestion. He notes that individual minds are enormously complex, and this complexity is compounded when we move to the level of minds interacting in a society. But, he says, we have a way of cutting through that complexity and apprehending, apparently directly, truths at the social level:

> The difficulties in knowing a single psychical entity are multiplied by the great varieties and uniqueness of these entities, by the way they work together in a society, by the complexity of natural conditions which bind them together, and by the sum total of mutual influences brought to bear in the succession of many generations which does not allow us to declare directly from human nature as we know it the state of affairs of earlier times or to infer present states of affairs from a general type of human nature. Nevertheless, all this is more than outweighed by the fact that I myself, who inwardly experience and know myself, am a member of this social body and that the other members are like me in kind and therefore likewise comprehensible to me in their inner being. (1883/1988, 98)

of evidence as a source of validation in the human or historical sciences but not within the natural ones, then the dyadic tradition would have succeeded in distinguishing the natural from the human sciences, but at the cost of crossing the boundary into metaphysics. This is because the special sort of perception, insofar as it is special and not ordinary public observation, will lead to the same sort of irresolvable (wearisome) controversies that metaphysics does.

Carnap does not in fact attack the human sciences as unempirical; he thinks they are perfectly legitimate sciences resting on the same sort of public evidence that other sciences do. He does not attack them as metaphysical. Rather, he poses a dilemma to the dyadic tradition: decide whether to claim that there is special evidence involved for the human sciences. If not, then the argument for the sharp distinction between two kinds of sciences falls apart. If the claim is that there is a special kind of perception in the human sciences, then they dyadic tradition convicts itself of having those sciences engage in irresolvable metaphysical controversies.

So in a sense, Carnap's response to the dyadic tradition is much the same as his mature response to metaphysics. We need clarity. State clearly what you take as evidence and as argument forms. And let us see where that leads.

It would be foolish to think that Carnap could "overcome" metaphysics in the sense that it would disappear. And it would be foolish to think that he could end the more that two hundred year old controversy over the relations between the human and natural sciences. But we can come to understand more nearly what is at stake in a controversy that has been going on for two hundred years, and that is an important step. And in the process we can get clearer about our own methods in history, in philosophy, and in science. And we can refine them as well. If we lay out those methods clearly, pursue, with an open mind, those methods we think most fruitful, and allow others to do the same, we can let history judge our progress.

Bibliography

Anderson, R. Lanier. 2012. The Debate Over the *Geisteswissenschaften* in German Philosophy. In *The Cambridge History of Philosophy 1870–1945*, ed. Thomas Baldwin. Cambridge: Cambridge University Press.

Berlin, Isaiah. 1996. *The Sense of Reality: Studies in Ideas and Their History*, ed. Henry Hardy. London: Chatto & Windus.

———. 1998. *The Roots of Romanticism*, ed. Henry Hardy. Princeton: Princeton University Press.

———. 2000. *Three Critics of the Enlightenment: Vico, Hamman, Herder*, ed. Henry Hardy. Princeton: Princeton University Press.

Carnap, Rudolf. 1928a/1967. *The Logical Construction of the World*. Trans. Rolf A. George. Berkeley: University of California Press.

———. 1928b/1967. *Pseudoproblems in Philosophy*. Trans. Rolf A. George. Berkeley: University of California Press.

———. 1932a/1959. The Elimination of Metaphysics Through the Logical Analysis of Language. Trans. Arthur Pap. In *Logical Positivism*, ed. A.J. Ayer, 60–81. New York: Free Press.

———. 1932b/1963. The Physical Language as the Universal Language of Science. Trans. Max Black (revised). In *Readings in Twentieth-Century Philosophy*, ed. William Alston and George Nakhnikian, 393–424. New York: Free Press.

———. 1932c/1959. Psychology in Physical Language. Trans. George Schick. In *Logical Positivism*, ed. A.J. Ayer, 165–198. New York: Free Press.
———. 1934/1937. *The Logical Syntax of Language*. Trans. Amethe Smeaton. London: Kegan Paul Trench, Trubner & Co.
———. 1938. Logical Foundation of the Unity of Science. In *Encyclopedia and Unified Science*, ed. Otto Neurath, et al. International Encyclopedia of Unified Science 1(1): 42–62. Chicago: University of Chicago Press.
———. 1963. Intellectual Autobiography. In *The Philosophy of Rudolf Carnap*, ed. Paul Arthur Schilpp. LaSalle: Open Court Publications.
Cassirer, Ernst. 1910/1923. *Substance and Function*. Trans. William Swabey and Marie Swabey. Chicago: Open Court Publishing Co.
———. 1929/1957. The Philosophy of Symbolic Forms, *Vol. 3, The Phenomenology of Knowledge*. Trans. Ralph Manheim. New Haven: Yale University Press.
Creath, Richard. 1996, The Unity of Science: Carnap, Neurath, and Beyond. In (Galison and Stump 1996), 158–69.
———. 2010. The Role of History in Science. *Journal of the History of Biology* 43: 207–214.
Damböck, Christian. 2012. Rudolf Carnap and Wilhelm Dilthey: 'German' Empiricism in the *Aufbau*. In *Rudolf Carnap and the Legacy of Logical Empiricism*, ed. Richard Creath. Dordrecht: Springer.
Dilthey, Wilhelm. 1883/1988. *Introduction to the Human Sciences: An Attempt to Lay a Foundation for the Study of Science and History*. Trans. Ramon J. Betanzos. Detroit: Wayne State University Press.
Friedman, Michael. 1987. Carnap's *Aufbau* Reconsidered. *Nous* 21: 521–545.
Gabriel, Gottfried. 2007. Carnap and Frege. In *Cambridge Companion to Carnap*, ed. Michael Friedman and Richard Creath, 65–80. Cambridge: Cambridge University Press.
Gallison, Peter, and David Stump. 1996. *The Disunity of Science: Boundaries, Contexts, and Power*. Stanford: Stanford University Press.
Hempel, Carl G. 1942. The Function of General Laws in History. *Journal of Philosophy* 39: 35–48.
Lilla, Mark. 1993. *G. B. Vico: The Making of an Anti-modern*. Cambridge, MA: Harvard University Press.
Mormann, Thomas. 2006. Werte bei Carnap. *Zeitschrift für.*
Naess, Arne. 1965/1968. *Four Modern Philosophers: Carnap, Wittgenstgein, Heidegger, Sartre*. Trans. Alistair Hannay. Chicago: University of Chicago Press.
Quine, W.V.O. 1951. Two Dogmas of Empiricism. *Philosophical Review* 60: 20–43.
———. 1971. Epistemology Naturalized. *Akten des XIV. Internationalen Kongresses für Philosophie*, 6:87–103.
Richardson, Alan. 1998. *Carnap's construction of the World: The Aufbau and the Emergence of Logical Empiricism*. Cambridge: Cambridge University Press.
Rickert, Heinrich. 1896–1902/1986. *The Limits of Concept Formation in Natural Science: A Logical Introduction to the Historical Sciences* (abridged ed.). Trans. and Ed. Guy Oakes. Cambridge: Cambridge University Press.
———. 1899/1962. *Science and History: A Critique of Positivist Epistemology*. Trans. George Reisman. Princeton: D. Van Nostrand Co.
Suppes, Patrick. 1978. The Plurality of Sciences. *PSA 1978, Vol 2*, ed. Peter Asquith and Ian Hacking. E. Lansing: Philosophy of Science Association.
Windelband, Wilhelm. 1894/1980. History and Natural Science. Trans. Guy Oakes. *History and Theory* 19: 169–185.
———. 1893/1901/1958. *A History of Philosophy, Vol. II: Renaissance, Enlightenment, and Modern*. Trans. and Rev. ed. James H. Tufts. New York: Harper & Row.

Chapter 2
Saving Models from Phenomena: A Cautionary Tale from Membrane and Cell Biology

Axel Gelfert and Jacob Mok

Abstract This paper investigates one of the great achievements of twentieth-century cell biology: determining the structure of the cell membrane. This case differs in important ways from the better-known case of the identification of the DNA double helix as the carrier of genetic information, especially regarding the evaluation of potential evidence in light of prior theoretical commitments. Whereas it has been argued that adherence to a structural hypothesis enabled Watson and Crick to ignore a surplus of (potentially confusing) empirical findings, similar adherence to an elegant and universal structural hypothesis, we argue, unduly shielded the so-called 'unit-membrane' model from legitimate challenges on the basis of known phenomena.

2.1 Introduction

Molecular cell biology has been the subject of a number of recent studies in the history and philosophy of biology.[1] But the focus of attention has been uneven: whereas the search for the physical basis of heredity—culminating in the discovery of the DNA double helix and its genetic code—has been explored in much detail, other major developments in twentieth-century cell biology have yet to be analyzed in detail. This paper discusses one such example: the identification of the structure of the cell membrane. As we shall argue, this example not only casts light on how empirical findings gain the status of evidence, and how recognition of this status can be obstructed by prior theoretical commitments, but it also makes for an interesting contrast case to the discovery of the DNA double helix. In the latter case, considerable uncertainty about the physical basis of heredity was followed by an almost immediate consensus about the DNA double helix as the carrier of

[1] Recent monographs include (Keller 2002), (Weber 2005), (Bechtel 2006), and many more.

A. Gelfert (✉) • J. Mok
Department of Philosophy, National University of Singapore,
3 Arts Link, 117570 Singapore, Singapore
e-mail: axel@gelfert.net

F. Stadler (ed.), *Integrated History and Philosophy of Science*, Vienna Circle Institute Yearbook 20, DOI 10.1007/978-3-319-53258-5_2

genetic information; by contrast, in the membrane case, the majority view that membranes consisted of a lipid bilayer, uniformly coated with globular proteins—which held sway for more than 30 years—was only gradually challenged once new experimental techniques (such as freeze-fracture electron microscopy) were adopted in the 1960s and the evidentiary significance of previous findings and mismatches was recognized.

The rest of this paper is organized as follows: Section 2.2 reviews the interplay between models, theory, and evidence in the life sciences and introduces a useful recent distinction between *signs* and *conditions* of evidentiary success. Section 2.3 gives a detailed reconstruction of the development of membrane models in cell biology, paying special attention to 'hits and misses' in the evaluation of potential evidence. Section 2.4, in the light of this case study, argues for the inclusion of experimental techniques and epistemological strategies of their practitioners among the conditions of evidentiary success. Section 2.4 concludes the paper by contrasting the—guiding, yet also constraining—role of theoretical commitments in evaluating potential evidence for the complementary cases of membrane models and Watson and Crick's DNA double helix model. Whereas in the latter case shielding the double helix model from prima facie disconfirming evidence turned out to have been a fortuitous shortcut to what we now take to be the best scientific account of the structure of DNA, the example of early membrane biology shows that trying to save a structurally appealing model from empirical challenges may lead to the tacit (and often unwarranted) dismissal of legitimate evidence, especially when such evidence is the result of new experimental techniques.

2.2 Models, Theory, and Evidence: The View from the Life Sciences

Well-entrenched debates in the philosophy of science—concerning the relation between theory and observation, data and hypothesis, confirmation and evidence—often take on a new level of complexity when applied to the life sciences. Methods and approaches that have been found to be suitable for the physical sciences—though themselves by no means simple—often resist straightforward application to the life sciences. As Nicolas Rasmussen puts it: 'This situation arises largely from the nonspecificity, incompleteness, and over simplicity of physical and chemical theory in its application to biological complexity: either theory is too general to be of real use, or when worked out in detail for a given problem, too uncertain and itself subject to revision.' (Rasmussen 1993, 231) At the level of experimentation, improvements in the available instruments and techniques—such as the shift from light to electron microscopy—not only open up novel realms of inquiry, but also raise new questions about the calibration of instruments, commensurability of empirical findings, and concordance (or discordance) of evidence. As biologists began to study the molecular basis of life, the line between what Ian Hacking (1983) has characterized as the contrast between 'representing' and 'intervening' by necessity had to be frequently crossed. As Evelyn Fox Keller puts it: 'Biologists could see more, but they could no longer watch' (2002, 214).

In spite of these challenges, a growing body of philosophical literature explores how models, theories, and evidence come together in experimentation in the life sciences.[2] Such studies have often taken the form of detailed case studies 'of scattered and usually quite recent episodes of experimentation', sometimes at the expense of 'diachronic studies that would reveal how [various general] guiding principles came to be applied, how the principles may have mutated in character over time, or how the principles may have changed in their mix or hierarchy relative to one another amongst the set deployed by experimenters in a field' (Rasmussen 1997, 12). While there is certainly a risk that merely assembling more and more case studies might result in historians and philosophers of science no longer being able to see the forest for the trees, nonetheless a well-chosen case study, in our opinion, may not only provide diachronic depth, but may also cast a spotlight on philosophical problems and challenges. Indeed, the case we wish to discuss in the next section—of models of the cell membrane—aims to achieve just that, and it does so by looking in depth at how the interplay of models, theoretical hypotheses, and empirical findings (and their interpretations) has developed over time, sometimes in response to advances in experimental techniques and instrumentation, sometimes by fitting evidence to a preferred theoretical hypothesis concerning the structure of cell membranes.

In what follows, special emphasis will be placed on how various experimental findings acquired evidentiary significance regarding the various theoretical hypotheses that were proposed for the structure of the cell membrane. At an abstract level, it will be useful to distinguish, following Jacob Stegenga (2013), between *signs of success* and *conditions of success* in relation to evidence in biology. Whereas the former refers to formal features (e.g. in terms of probabilistic support) of statements of evidence once an empirical finding has been recognized as such (rather than, say, dismissed as background noise), the latter refers to features of methods (including prior to the consideration of any evidence generated by it) and of the evidence itself—e.g. recurring patterns, concordance from multiple streams of evidence, plausibility and so forth. Stegenga illustrates this point with an analogy: when we try to assess whether a particular wine is good, we might rely on the fact that it has been awarded 90 out of 100 points by an expert wine critic or, alternatively, we might appeal to features of the wine itself (its bouqet and taste, say) or to specific aspects of its method of production (e.g., where it was grown).[3] The first type of approach relies on identifying purely formal *signs of success*, whereas assessments of the second type engage with substantive *conditions of success*—considerations of the sort a wine critic might herself rely on in assigning a numerical score. Any approach—whether in oenology or the philosophy of science—that places signs of

[2] See e.g. (Creath and Maienschein 2000), (Rheinberger 1997), and refs. in fn. 1.; for a review of the philosophical literature on scientific models, see (Gelfert 2016).

[3] See (Stegenga 2013, 982).

success at the heart of its project may, in this sense, be considered necessarily *post hoc*, since 'one already must possess and have evaluated the evidence with the conditions of success in order to do business in the signs of success tradition' (Stegenga 2013, 982).

Taking as one of his case studies Avery et al.'s successful identification, in the 1940s, of nucleic acids as the material basis of heredity, Stegenga argues—plausibly, in our judgment—that 'characterization of evidence in terms of the conditions of success has the virtue of accurately describing those aspects of evidence which appear to matter to scientists' (Stegenga 2013, 1002), not least the quality and relevance of the methods employed, and the believability, patterns, and concordance of the empirical evidence thus generated (ibid., 997). Shifting the emphasis in philosophical discussions of scientific evidence from its formal features to its diverse conditions of success has the further advantage of being able to account for the fact that, when dealing with new methods and technologies of producing data, 'rather than simply assessing features of the method, scientists also assess features of the evidence itself' (Stegenga 2013, 988).[4] Yet, as Jane Maienschein has argued, debates in science often concern 'what is to count as evidence or how much certain evidence is to count for or against a given argument' (Maienschein 2000: 122), and so we should be prepared to find that judgments concerning the evidentiary significance of empirical results will often themselves be driven by (potentially diverging) epistemological strategies.

2.3 The Case of Membrane Models in Cell Biology

Though the idea that the smallest units of life might be 'envelope-bounded vesicles' predates the nineteenth century, the first systematic experiments probing cell membrane structure were performed by Charles Overton only in 1895.[5] Overton discovered that non-polar (lipid-soluble) substances—Overton was using olive oil—were entering the cell faster than substances that would not mix with lipids. This led Overton to hypothesize that non-polar molecules might pass through the membrane by 'dissolving' in the membrane's 'lipid interior'. Thirty years later, Evert Gorter and François Grendel (1925) extracted the contents of red blood cells, leaving only the membranes. Analysis of these remnants, called erythrocyte ghosts, revealed a lipid presence in the membranes themselves. Furthermore, these lipids were found to possess a surface area twice that of the red blood cells used in the experiment. Repeated studies with erythrocytes from other animals yielded similar results. Gorter and Grendel concluded that erythrocytes 'are covered by a layer of fatty substances two molecules thick', suggesting that biological membranes consisted of a bimolecular lipid layer.

[4] On this point, see also (Bechtel 2000).

[5] See e.g. (Bechtel 2006, 77–80).

Subsequent studies indicated that lipids could not be all there is to membrane structure. Experiments measuring the surface tension of starfish egg membranes were carried out, by determining the force necessary to compress an egg between two slide cover slips (see Cole 1932). The values obtained were in the region of 0.1 dyne/cm and matched those derived from measuring the surface tension of oil droplets within cells, using a centrifuge microscope. (See Harvey and Shapiro 1934.) In both cases, the values revealed surface tensions far below those found in pure lipids. This was suspected to be the result of a protein presence and subsequent experiments with albumen and benzene seemed to confirm this—when the two were mixed with water, protein molecules from the albumen were discovered to have aligned themselves at the benzene-water interface, resulting in a lower surface tension than at a benzene-water interface alone.[6] This protein presence was corroborated a few years later when it was found that the rotation of polarized light by erythrocyte ghosts could only be explained by a high protein concentration in erythrocyte membranes.[7] Such results, and a general fascination with the properties of polypeptides—which held out the promise of explaining the 'sieve-like' function of cell membranes—led some researchers to also consider predominantly protein-based models of the cell membrane. However, as Runar Collander showed in a series of experiments with membranes that had been prepared by fixing gelatine with formaldehyde, such protein membranes did 'not in the slightest' show the observed preferential permeability to lipoid-soluble substances. This, Collander concluded, was 'an indirect argument in favour of the view that one could hardly explain the permeability properties [of living cells] without assuming a lipoid presence' (Collander 1927: 221; our translation).

Based on these findings and precursor models, Hugh Davson and James Danielli, in 1935, proposed a model that comprehensively accounts for the presence of both lipids and proteins and, in this sense, constitutes the first model of biological membranes as *lipid-protein structures*. Specifically, Davson and Danielli suggested that cellular membranes consisted of a sandwich of amphipathic lipids (i.e. molecules with both hydrophilic and hydrophobic ends), arranged in a bilayer with the polar hydrophilic heads of the lipid molecules pointing outwards, which are then capped by protein monolayers (see Fig. 2.1, from Danielli and Davson 1935). Though influenced by the aforementioned discoveries, none of the earlier findings can account for, let alone necessitates, the *particular configuration* of protein molecules in the Davson-Danielli model. In his biography, Danielli later admits that this configuration was motivated largely by a single experiment conducted just a year earlier.[8] During his doctoral research, Danielli had discovered that oil droplets obtained from mackerel eggs injected into a protein solution would become crenated—the protein molecules were naturally adhering to the oil droplet micelles. (See Danielli and Harvey 1934). Danielli's model of protein layers forming weak interactions with the polar heads of a membrane's lipid bilayer seemed to fit neatly with the

[6] See (Danielli 1938).

[7] See (Schmitt et al. 1936), (Schmitt 1938).

[8] See (Stein 1986).

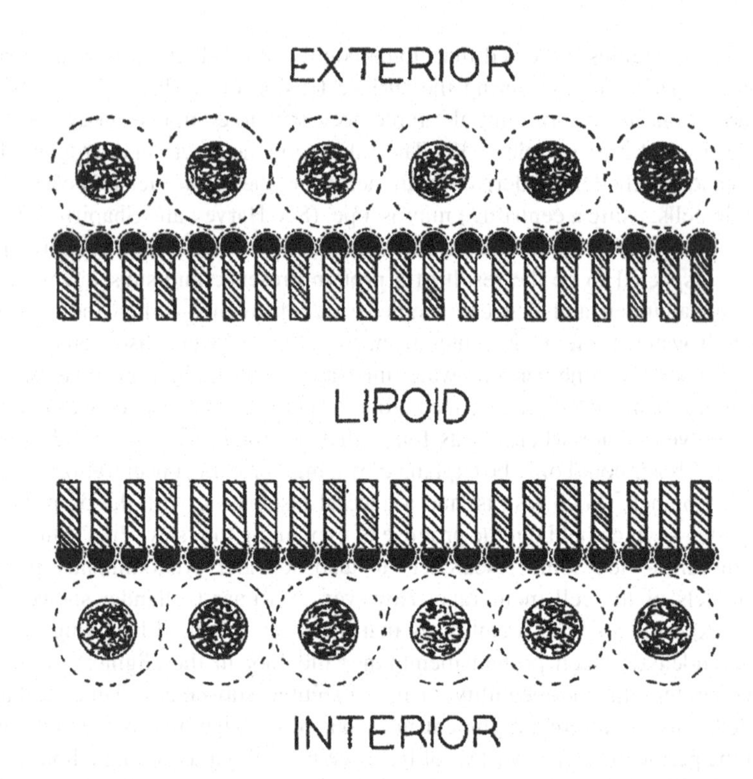

Fig. 2.1 The first diagram of the Davson-Danielli model published in 1935. The polar heads of the molecules in the lipoid core are oriented outwards which are then weakly bonded to a layer of globular proteins (Reproduced with permission from the publisher)

ready adsorption of protein molecules to lipid micelles. This 'neat fit' was seemingly all it took for the model to become widely promoted and accepted in the 1930s.

The advent of the electron microscope in the 1950s yielded what was thought to be further 'proof' of the Davson-Danielli model. Early electron micrographs of erythrocyte membranes revealed a trilaminar structure consisting of a light band approximately 50 Å thick sandwiched by two dark bands each approximately 25 Å thick (see Fig. 2.2). Most biologists assumed that the heavy-metal colouring used adhered only to the protein layer and the hydrophilic lipid head and would not stain the hydrophobic lipid tails, with the presumed protein-lipid-protein structure accounting for the dark-light-dark banding. Similar examination of other cell types and cell organelles yielded the same banding patterns, leading many to infer that the Davson-Danielli model was adequate as a model of *all* biological membranes.[9] In short, the electron micrographs and the membrane model became increasingly associated as explanations for each other. But this circularity did not diminish the

[9] For a review by one of the main contributors to electron microscopy on different membrane types, see (Robertson 1963).

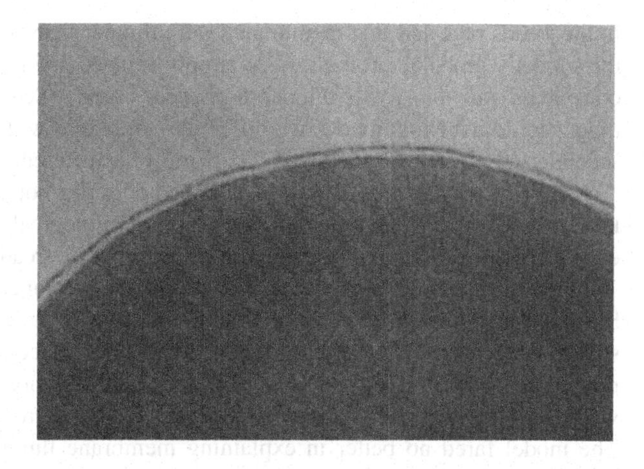

Fig. 2.2 Electron micrograph of a human erythrocyte membrane consisting of two dense strata separated by a light band. The overall thickness of the membrane bands is about 130–140 Å. (©1981 Robertson, The Journal of Cell Biology. 91: 189s–204s; reproduced with kind permission from Rockefeller University Press)

model's appeal, much of which derived from its purported global applicability to all membranes: it presented a unified account of membrane structure and became widely known as the 'unit-membrane model'—all biological membranes were believed to be composed of Davson-Danielli-type protein-lipid-protein sub-units.[10]

Despite its less than rigorous motivation, the 'unit-membrane' model became *the* definitive model of membrane structure: up till the early 1970s, textbooks regularly described biological membranes as 'lipoprotein sandwiches' and included electron micrographs that were presented as proof of the model.[11] This ready acceptance and the more than 30-year persistence of the model is all the more perplexing when considering that it could not account for well-established membrane properties and even contradicted some of the relevant experimental knowledge about membrane behaviour. To begin with, it was already known before the creation of the model that membrane proteins are amphipathic and not completely polar (i.e., hydrophilic).[12] The hydrophobic regions of such protein molecules would avoid contact with polar substances like water and the polar heads of membrane lipids, making the formation of a protein cap as seen in the model highly unlikely. Notably, this amphipathic character of membrane proteins renders Danielli's previously mentioned experiment with *polar* proteins and mackerel oil (!) all but irrelevant to biological membranes in a largely aquatic medium.

In addition, biologists were already aware of the functional variation between membranes (for example between mitochondrial, cellular, and Golgi apparatus membranes). The idea that all membranes share the same structure, while appealing, did not account for the different roles played by different cell and organelle membranes. The problem became even more salient when better staining techniques

[10] This is not to deny that there are complications, such as the existence of cell walls in plants, i.e. of carbohydrate-protein complexes formed through secretion between neighbouring plant cells. For a (somewhat dated) historical sketch, see (Smith 1962).

[11] See, e.g. (Yost 1972).

[12] See (Mudd and Mudd 1926).

in the 1960s revealed that membranes were thinner than first thought (<80 Å) and the model's globular proteins were simply replaced—in move that may well be considered *ad hoc*—with thinner β-pleated sheets. (See Robertson 1959). The model could not explain how two different membranes with proteins with the same secondary β-pleated structure could perform very different functions.

Most importantly, the Davson-Danielli model did not account for the already known lipid-soluble and fluidic properties of biological membranes. Overton's early experiments had revealed that membranes were readily permeable to lipid-soluble molecules. Yet how could such non-polar substances pass through a membrane if it was protected by a polar protein sheet like in the proposed model? Danielli subsequently (1954) altered his model to include, at certain intervals along the membrane, protein-lined pores to facilitate such diffusion. Such pores, however, would still be lined with *hydrophilic* proteins that would repel non-polar molecules. The model fared no better in explaining membrane fluidity: biologists had long observed large single-celled organisms form food and excretory vacuoles—something that seemed difficult with membranes with a β-protein cap. Prior observations of membrane interactions with microsurgical tools—membranes would bend around probes and, if broken, reseal themselves—were likewise impossible to conceive of in the Davson-Danielli model; the rigid protein layers postulated in the model could not accommodate such membrane fluidity without permanent bond-breaking or membrane-tearing.

It was not until more than thirty years after Davson and Danielli's first proposal, that the unit-membrane model was being seriously reconsidered in the 1960s. Previously ignored explanatory shortcomings of the model, such as its inability to account for the functional variability of membranes in different parts of the cell, were now attracting the attention of researchers. New empirical findings, too, put pressure on the validity of the unit-membrane model. (For a concise historical discussion of this transition, see Morange 2013). Thus, the existence in the cell membrane of 'subunits', which appeared in the form of small particle-like inclusions, was revealed by advances in electron microscopy; however, it remained unclear whether these were evidence of a more complex substructure of the membrane itself or whether they were simply the result of fixation and staining procedures that physically altered the samples. With the development of freeze-etching electron microscopy by Daniel Branton in 1966, which 'made it possible to prepare biological specimens for electron microscopy without subjecting them to the usual chemical fixatives and embedments required in other preparatory techniques' (Branton 1979, 9), it became clear that the inclusions—'subunits' or 'particles'—found in membranes could no longer be brushed aside as mere 'artifacts'. Increasing insight into the dynamic properties of proteins, including their ability to self-assemble, gradually weakened belief in the uniformity of the membrane as a lipid bilayer sandwiched between two protein sheaths. Michel Morange describes the late 1960s as a transition period during which 'there coexisted different models of membranes that shared the same principle: the structure of membranes resulted from the structure of their proteins and from the interactions of these proteins' (2013, 4)—yet the details of this structure and interaction remained sketchy.

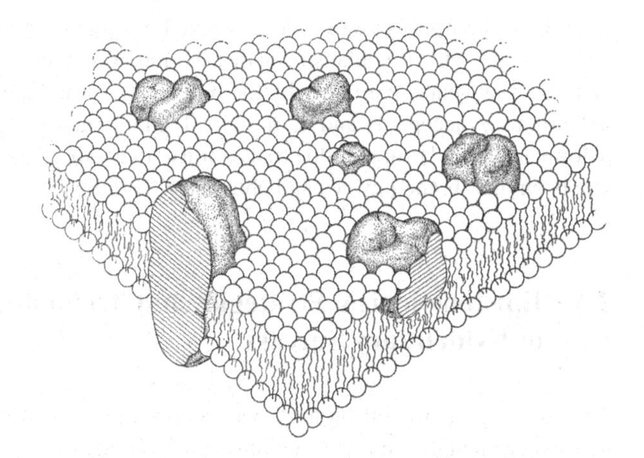

Fig. 2.3 'The lipid-globular protein mosaic model with a lipid matrix (the fluid mosaic model)' (From Singer and Nicolson 1972: 723. Reprinted with permission from AAAS)

What is generally considered to have sealed the fate of the unit-membrane model was the observation that proteins could readily diffuse laterally *within* membranes. Hubbell and McConnell (1969) studied the motion of spin-labelled phospholipids in membranes, demonstrating a greater molecular mobility than expected. Other researchers (Frye and Edidin 1970) analyzed the behaviour of 'fused mouse and human cells and showed by indirect immunofluorescence that the antigens present at the surface of the two cells were rapidly mixed in the membrane of the heterokaryon' (Morange 2013, 4). As Morange puts it, such results were 'compatible with the lipid bilayer model of the membrane, but incompatible with models proposing that membranes were organized in a quasi-crystalline state by interactions between their protein constituents' (ibid.). Frye and Edidin's study was particularly influential: 'the quality of the images obtained by Frye and Edidin, their resolution in time, as well as the abundance of controls that they performed to eliminate other possible explanations for the rapid intermixing of antigens gave their experiment a huge impact' (ibid.).

By the early 1970s, the diversity of findings had rendered the rigid, highly-ordered unit-membrane model untenable. As an alternative model—which remains valid today, with only minor modifications—the 'fluid mosaic model' was proposed by S. Jonathan Singer in 1971. It distinguishes between (globular) integral and peripheral membrane proteins, based on whether they are loosely attached to, or penetrate through, the lipid-layer system of the membrane. (See Fig. 2.3.) Importantly, proteins 'at long range are randomly distributed in the plane of the membrane' (Singer and Nicolson 1972, 723). By giving up on the idea of a highly-ordered 'unit-membrane', the fluid-mosaic model is able to accommodate experimental findings that had until then been deemed contradictory—among them the confusing finding that 'different types of experiments suggest strong protein-lipid interactions on the one hand, and weak or no interactions on the other': as Singer and Nicolson argue, this apparent 'paradox' can be resolved 'in a manner consistent with all the data if it is proposed that, while the largest portion of the phospholipid is in bilayer form and not strongly coupled to proteins in the membrane, a small fraction of the lipid is more tightly coupled to protein' (Singer and Nicolson 1972, 722–3)—due to the irregular spatial distribution of proteins throughout the membrane. Within a few years—certainly by the mid-1970s—the 'new model was rapidly and unanimously accepted, and it remained unaltered during the next forty years' (Morange 2013, 3), largely due to its ability 'to capture and integrate diverse experiments on membrane physics and chemistry' (Edidin 2003, 416).

2.4 Epistemological Strategies and Technological Conditions of Evidentiary Success

The rise of modern cell biology was deeply intertwined with the development of new experimental techniques and instruments that probed the inner workings of the cell. As the histologist Ramón y Cajal put it in somewhat purple prose, 'every advance in staining technique is something like the acqusition of a new sense directed towards the unknown' (quoted after Keller 2002, 215). Novel experimental techniques established themselves alongside more traditional methods such as optical microscopy; this required a continuous reassessment of how results from different methods could be made commensurable and could be reconciled.

Adopting a bird's-eye view of the rapid growth of cell biology in the light of electron microscopy and cell fractionation, Bechtel writes of the two methods:

> Both provided patterns of results that were determinate and repeatable. Each secured consilience of the results it generated with those obtained in other ways and, especially, with each other. (Bechtel 2006, 160)

Such consilience and mutual support across different methods, however, was not always initially apparent to their practitioners, many of whom brought divergent 'interpretive modes' (Rasmussen 1997, 124) and epistemological strategies to bear on the same results. Indeed, as Jane Maienschein has noted, it is often 'epistemic convictions that dictate what will count as acceptable practice and how theory and practice should work together to yield legitimate scientific knowledge' (Maienschein 2000, 123). As an example, consider Stuart Mudd, one of the contributors to cell membrane research (see fn. 12), who 'worked to weaken the overall epistemological authority of light microscopy' (Rasmussen 1997, 99); likewise, even among those who embraced the new techniques, there persisted methodological disputes concerning the uses to which the newly generated evidence—in particular, electron micrographs—should be put.

One such dispute, which has been given some attention by historians and philosophers of cell biology (Rasmussen 1997, 124–149; Bechtel 2006, 155–156), involved George Palade (at the Rockefeller Institute) and Fritiof Sjöstrand (at Stockholm's Karolinska Institute), both of whom were researching the mitochondrion. Whereas Palade's use of electron micrographs 'was primarily qualitative—to provide morphological perspective on the biochemical information generated [through other methods]' (Bechtel 2006, 155), Sjöstrand's approach gave pride of place to the quantitative findings of electron microscopy, e.g. concerning the size of cellular components, without attempting to subordinate them to prior biochemical theories. Thus, whereas Palade refrained 'from any effort to visualize [... the] membrane architectures' of mitochondria, Sjöstrand argued 'that the thickness and appearance correspond to [...] the then-current "sandwich" theory of membrane biochemistry, from two layers of protein coating a lipid bilayer' (Rasmussen 1997, 131/129)—that is, were similar in kind to the cell's outer membrane structure.

While such methodological disputes are of intrinsic historical interest, they are also of broader philosophical significance, insofar as they raise 'doubts about the degree of technological determinism that can be attributed to experimental systems'; as Rasmussen notes, in the particular case at hand 'the two camps had practices that were so similar technically, yet they used the results so differently that they came to conflicting conclusions' (1997, 125). Instead of technological determinism, we propose to speak of *technological conditions of evidentiary success*—that is, conditions of success (as discussed in Sect. 2.2) which are associated with specific technologies of experimentation. The mere availability of experimental techniques is never sufficient to bring about lasting theory change (in this case, concerning theories of membrane structure), but always requires conducive epistemological strategies among their practitioners. For example, Branton reports that his freeze-etching method—now credited with paving the way for the fluid-mosaic model—was met with scepticism, even from his former supervisor, Hans Moor.

From early on, it was clear 'that freeze-etching frequently exposed vast expanses of a cell's membranes to inspection in the electron microscope',[13] yet whereas Moor 'interpreted these expanses as the surface of the various membrane systems' themselves, Branton came to realize that this interpretation 'could not be reconciled with the known surface properties' of membranes. Instead, 'the fracture process used in freeze-etching was splitting biological membranes', which led to a significant reinterpretation of the empirical findings. One of the main outcomes was the demonstration that, although 'the membrane continuum was composed of a bilayer', the reinterpreted findings 'explicitly denied the notion of a biological membrane that was spatially uniform'. Without the advent of freeze-etching technology, the spatial non-uniformity in the samples might never have been observed. But, even more importantly, without the hard work of establishing shared standards of veridical interpretation for the new technique—which, as Branton's example shows, oftens involves going against seemingly well-established prior theories and findings—the observed non-uniformities might have been easily brushed off as due to contamination or 'noise'. Finally, the new technique brought into sharp focus previously unrecognized problems with the previous interpretation, which turned out to have been 'based on a set of contradictory assumptions'; this suggests that certain technological conditions must indeed be met for empirical findings to reveal their true evidentiary significance.

Acknowledging that experimental techniques contribute to, but do not solely determine, conditions of evidentiary success also coheres well with the observation

[13] All direct quotes from (Branton 1979: 9).

that what made some experiments in membrane research so influential were precisely those features (quality of the evidence, cohesion across time-resolved observations, determinate and repeatable patterns of results, etc.) that are also characteristic of evidentiary success in other domains. Consider Frye and Edidin's demonstration of rapid mixing of proteins in heterokaryons. Widely credited with clinching the case for more 'fluid' accounts of the cell membrane, Frye and Edidin's experiments displayed cohesion across different types of experimental techniques (use of antigen markers, inhibiting the cell's metabolism), produced visually salient evidence through the use of immunofluorescence, and lent themselves to successful replication and modification.[14]

2.5 Theoretical Commitments as Guide and Constraint

Disputes of the type discussed in the previous section, concerning novel experimental techniques, are of course not unusual in 'micro histories' of scientific developments. What makes the case of membrane models in cell biology (Sect. 2.3) significant, is the way it illustrates the complex relation, across a span of several decades, between theoretical commitments and—old and new—empirical evidence.

In another recent case study of the interplay between model, theory, and evidence in molecular biology, Samuel Schindler argues that Francis Crick and James Watson's discovery of the structure of DNA would not have been possible, had they not single-mindedly—and in the face of contravening evidence—pursued their double-helix hypothesis. Whereas 'Crick and Watson stuck to *one* structural hypothesis throughout', Rosalind Franklyn 'conducted experiments and, on the basis of the obtained evidence, she tried to infer the structure that would fit her data best' (Schindler 2008, 627)—alas, her attempt 'to use her experimental data as fully as possible' (as Crick described it; 1988, 68) turned out to be self-defeating, given the diverse range of X-ray photographs, some of which seemed to contradict any helical structure. From the perceived superiority of the 'top-down' strategy of *sticking with a structural hypothesis*, Schindler draws far-reaching conclusions, ruling out both 'the abductive component of IBE' and 'classical hypothetical-deductive confirmation' as ways of describing this historical episode, and concludes that 'taking particular evidence too seriously blocked some researchers from making discoveries, which were made by others who ignored the same evidence' (2008, 651). Sometimes, then, saving models from empirical challenges—as in the case of Watson and Crick sticking with the structurally appealing DNA double helix model, even in the face of seemingly disconfirming X-ray diffraction images—may turn out, by hindsight, to have been a fortuitous shortcut to what we now take to be the best scientific account of the underlying structure of reality.

[14] See (Frye and Edidin 1970).

However, the case of membrane models should give defenders of the 'top-down' strategy cause to pause. Unlike in the DNA double helix case, stubborn adherence to an elegant, simple, and universal (i.e., 'neat') structural hypothesis—in the form of the unit-membrane model, consisting of a lipid bilayer coated uniformly with proteins—did not advance but, on the contrary, hindered the discovery of the true structure of the cell membrane, which turned out to be a messy mosaic of different kinds of proteins embedded in, and attached to, a lipid bilayer without any discernible long-range order. Importantly, it was not only new phenomena such as the rapid intermixing of proteins in heterokaryons that should have called the 'elegant' unit-membrane model into question: recall that, as early as the late 1920s, membrane proteins were known to be amphipathic and not completely polar and, arguably, should never have been considered obvious candidates for forming a uniform protein cap in the first place (see Sect. 2.3). The very theoretical virtues of the unit-membrane model—its simplicity and assumed universality, combined with the cognitive attractiveness of its easy visualizability—shielded it from potentially challenging phenomena, whose evidentiary significance was not realized until new technological developments and experimental techniques created the conditions that made a successful synthesis in the form of the fluid-mosaic model possible. Whenever the 'beauty', 'elegance', or 'universality' of a model are adduced as a reason for saving it from empirical challenges, caution is therefore advised. What matters is that incoming evidence be assessed on the basis of a range of criteria—which must include technological conditions of evidentiary significance that reflect the most current and most reliable techniques available.

References

Bechtel, William. 2000. From Imaging to Believing: Epistemic Issues in Generating Biological Data. In *Biology and Epistemology*, ed. Richard Creath and Jane Maienschein, 138–163. Cambridge: Cambridge University Press.

———. 2006. *Discovering Cell Mechanisms: The Creation of Modern Cell Biology*. Cambridge: Cambridge University Press.

Branton, Daniel. 1979. This Week's Citation Classic. *Current Contents: Agriculture, Biology & Environmental Sciences* 1: 9.

Cole, Kenneth S. 1932. Surface Forces of the Arbacia Egg. *Journal of Cellular and Comparative Physiology* 1 (1): 1–9.

Collander, Runar. 1927. Einige Permeabilitätsversuche mit Gelatinemembranen. *Protoplasma* 3 (1): 213–222.

Creath, Richard, and Jane Maienschein, eds. 2000. *Biology and Epistemology*. Cambridge: Cambridge University Press.

Crick, Francis. 1988. *What Mad Pursuit: A Personal View of Scientific Discovery*. New York: Basic Books.

Danielli, James. 1938. Protein Films at the Oil-Water Interface. *Cold Spring Harbor Symposium on Quantitative Biology* 6: 190–195.

———. 1954. The Present Position of Facilitated Diffusion and Selective Active Transport. *Recent Developments in Cell Physiology, Colston Papers* 7 (1): 1–14.

Danielli, James, and Hugh Davson. 1935. A Contribution to the Theory of Permeability of Thin Films. *Journal of Cellular and Comparative Physiology* 5 (4): 495–508.

Danielli, James, and E. Newton Harvey. 1934. The Tension at the Surface of Mackerel Egg Oil, With Remarks on the Nature of the Cell Surface. *Journal of Cellular and Comparative Physiology* 5 (4): 483–494.
Edidin, Michael. 2003. Lipids on the Frontier: A Century of Cell-Membrane Bilayers. *Nature Reviews Molecular Cell Biology* 4 (5): 414–418.
Frye, Larry D., and Michael Edidin. 1970. The Rapid Intermixing of Cell Surface Antigens After Formation of Mouse-Human Heterokaryons. *Journal of Cell Science* 7 (2): 319–335.
Gelfert, Axel. 2016. *How to Do Science with Models: A Philosophical Primer*. Heidelberg/New York: Springer.
Gorter, Evert, and François Grendel. 1925. On Bimolecular Layers of Lipoids on the Chromocytes of the Blood. *Journal of Experimental Medicine* 41 (4): 439–443.
Hacking, Ian. 1983. *Representing and Intervening*. Cambridge: Cambridge University Press.
Harvey, E.N., and Herbert Shapiro. 1934. The Interfacial Tension Between Oil and Protoplasm Within Living Cells. *The Journal of Cellular and Comparative Physiology* 5 (2): 255–267.
Hubbell, Wayne L., and Harden M. McConnell. 1969. Motion of Steroid Spin Labels in Membranes. *Proceedings of the National Academy of Sciences* 63 (1): 16–22.
Keller, Evelyn Fox. 2002. *Making Sense of Life*. Cambridge, MA: Harvard University Press.
Maienschein, Jane. 2000. Competing Epistemologies and Developmental Biology. In *Biology and Epistemology*, ed. Richard Creath and Jane Maienschein, 122–137. Cambridge: Cambridge University Press.
Morange, Michel. 2013. What History Tells Us XXX. The Emergence of the Fluid Mosaic Model of Membranes. *Journal of Biosciences* 38 (1): 3–7.
Mudd, Stuart, and Emily B.H. Mudd. 1926. On the Surface Composition of Normal and Sensitised Mammalian Blood Cells. *Journal of Experimental Medicine* 43 (1): 127–142.
Rasmussen, Nicolas. 1993. Facts, Artifacts, and Mesosomes: Practicing Epistemology with the Electron Microscope. *Studies in History and Philosophy of Science* 24 (2): 227–265.
———. 1997. *Picture Control: The Electron Microscope and the Transformation of Biology in America, 1940–1960*. Stanford: Stanford University Press.
Rheinberger, Hans-Jörg. 1997. Experimental Complexity in Biology: Some Epistemological and Historical Remarks. *Philosophy of Science* 64: S245–S254.
Robertson, J.D. 1959. The Ultrastructure of Cell Membranes and Their Derivatives. *Biochemical Society Symposium* 16: 3–43.
———. 1963. Unit Membranes: A Review. In *Proceedings of the 22nd Symposium of the Society for the Study of Development and Growth*, ed. Michael Locke, 1–8. New York: Academic.
———. 1981. Membrane Structure. *Journal of Cell Biology* 91 (3): 189–204.
Schindler, Samuel. 2008. Model, Theory, and Evidence in the Discovery of the DNA Structure. *British Journal for the Philosophy of Science* 59 (4): 619–658.
Schmitt, Francis Otto, R.S. Bear, and Eric Ponder. 1936. Optical Properties of the Red Cell Membrane. *Journal of Cellular and Comparative Physiology* 9 (1): 89–92.
———. 1938. The Red Cell Envelope Considered as a Wiener Mixed Body. *Journal of Cellular and Comparative Physiology* 11 (2): 309–313.
Singer, Seymour J., and Garth L. Nicolson. 1972. The Fluid Mosaic Model of the Structure of Cell Membranes. *Science* 175 (4023): 720–731.
Smith, Homer W. 1962. The Plasma Membrane, with Notes on the History of Botany. *Circulation* 26 (5): 987–1012.
Stegenga, Jacob. 2013. Evidence in Biology and the Conditions of Success. *Biology & Philosophy* 28 (6): 981–1004.
Stein, Wilfred D. 1986. James Frederic Danielli: 13 November 1911–22 April 1984. *Biographical Memoirs of Fellows of the Royal Society* 32: 117–135.
Weber, Marcel. 2005. *Philosophy of Experimental Biology*. Cambridge: Cambridge University Press.
Yost, Henry Thomas. 1972. *Cellular Physiology*. Prentice-Hall: Englewood Cliffs.

Chapter 3
My Touchstone Puzzles. W.D. Hamilton's Work on Social Wasps in the 1960s

Guido Caniglia

> *Detailed data without concepts have little significance; and concepts without information can become vacuous fantasies. So, there has to be a continuous interweaving of fact and theory, both within a discipline and in the mind of individual scientists*
>
> (M.J. West-Eberhard)

3.1 Introduction

W.D. Hamilton, one of the most influential evolutionary biologists of the twentieth century, was fascinated and puzzled by social wasps. Towards the end of his academic career, Hamilton openly admitted that: "... it was to social life that wasps were providing my touchstone puzzles" (Hamilton 1996, vi). This article deals with Hamilton's attempts to understand social life in wasps mostly in the 1960s. First, it provides an overview of the reasons why Hamilton thought social wasps constituted a puzzle. Second, it shows how Hamilton tried to deal with this puzzle, by performing experiments, conducting observations and eventually modifying his theory by including factors, most notably inbreeding, that he had not clearly considered in its first formulation.

In *The Genetical Evolution of Social Behavior I & II*, Hamilton presented his theory of inclusive fitness addressing how genetic relatedness of individuals in a population affects the evolution of social behaviors (Hamilton 1964a, b). According to Hamilton, inclusive fitness theory was "applicable to social behavior under relatedness for any group in the living world" (Hamilton 1996, 20). The 1964 article is divided in two main parts. In Part I, Hamilton presented the "genetical mathematical model" of inclusive fitness theory (Hamilton 1964b, 2), which "allows for interactions between relatives on one another's fitness" (Hamilton 1964a, 1). Part II shows how the model applies to concrete biological cases and presents a wide variety of behaviors ranging from warning behaviors to parental care and parasitism

G. Caniglia (✉)
Leuphana Universität Lüneburg, Lüneburg, Germany
e-mail: guido.caniglia@leuphana.de

F. Stadler (ed.), *Integrated History and Philosophy of Science*, Vienna Circle Institute Yearbook 20, DOI 10.1007/978-3-319-53258-5_3

(Hamilton 1964b). With these examples, Hamilton attempted to discuss "whether there is evidence that it [the theory] does work effectively in nature" (1964b, 17).

Altruistic behaviors, say those behaviors that are costly to the individuals performing them and beneficial to the ones receiving them, do not give the actor of the behavior advantages in terms of fitness. On the contrary, they lower their fitness. Therefore, altruistic behaviors pose a special challenge for evolutionary explanations of social life (Hamilton 1963). Hamilton asked under what conditions can the genes underpinning such behaviors spread in a population (Hamilton 1963, 1964a). The theory of inclusive fitness, if applied to the evolution of biological altruism, describes the conditions favoring the increase in frequency of a gene with altruistic effects in a population. According to this theory, altruistic behaviors can actually spread, if the benefits of such behaviors fall on individuals who are genetically related to the actor of the behavior rather than on random members of a population (Hamilton 1963).

An important case of biological altruism is represented by the self-sacrificing behaviors of workers in colonies of social insects of the order Hymenoptera, wasps, ants and bees (Hamilton 1963). In Hymenoptera, workers give up their reproductive capacities in favor of one or few individuals that are able to reproduce (Hamilton 1963). In Part II, Hamilton famously presented a hypothesis about how inclusive fitness might apply to the special case of social Hymenoptera, characterized by the unusual sex determination mechanism called haplodiploidy. This hypothesis became famously known as the haploidiploidy hypothesis In dealing with altruism in Hymenoptera, Hamilton paid special attention to "cases which appear anomalous" (Hamilton 1964b, 17) and are not easily explained by the haplodiploidy hypothesis. Both primitively and highly social species of wasps showed two kinds of anomalous and puzzling behaviors (Hamilton 1964b). First, they contain multiple egg-laying queens. This phenomenon is usually known as polygyny, or pleometrosis. Second, they mate multiple times with different males. This phenomenon is usually known as polyandry. Both polyandry and polygyny tend to lower the relatedness of the individuals in a colony and, in doing so, pose serious challenges to explanations in terms of inclusive fitness (Hamilton 1964b).

Already in 1964, Hamilton surmised that inbreeding might play a central role in explaining wasps puzzling behaviors. He also hypothesized that inbreeding depends on the high viscosity of wasp populations in temperate climates, but above all in the tropics (Hamilton 1964b). Yet, only in the paper *Altruism and Related Phenomena mostly in Social Insects* (Hamilton 1972), after observing and experimenting on many wasp colonies during two trips to South America in 1963–1964 and in 1968–1969 and many conversations with scholars of wasps and evolutionary biologists, Hamilton updated the mathematical model of inclusive fitness presented in 1964, reassessed the value of his haplodiploidy hypothesis and newly discussed the puzzling behavior of wasp societies in the light of these ideas (Hamilton 1972; Segerstrale 2013). Yet, even with these adjustments, Hamilton was not able to solve the puzzle posed to his theory by polygyny and polyandry (Hamilton 1972).

This paper, first, briefly presents Hamilton's inclusive fitness theory as well as how it applies to the case of Hymenoptera with their special sex determination mechanisms. After focusing on the reasons and context of Hamilton's trip to South America, it explains why polygyny and polyandry make wasp societies a puzzle, or even an anomaly, for Hamilton's haplodiploidy hypothesis. Then, it gives an overview of how Hamilton addressed this puzzle both in his observations and experiments during his trips to Brazil as well as by reassessing his mathematical model of inclusive fitness theory in the 1972 paper. Finally, it concludes by presenting Hamilton's reflections about the puzzle again in 1972.

3.2 Inclusive Fitness Theory and the Haplodiploidy Hypothesis

Hamilton's inclusive fitness theory, the 'genetical mathematical model' presented in the 1964 article, explained the conditions favoring the increase in frequency of a gene with altruistic effects in a population (Hamilton 1964a). The so-called Hamilton's rule roughly summarizes Hamilton's ideas about the conditions favoring this process (Charnov 1977). In its condensed version, the formula shows that altruism can evolve if: br > c, where *r* is a measure of the degree of relatedness, the fraction of genes shared by the altruistic actor (let's say a worker wasp) and the recipient of the altruistic action (let's say, the brood of the queen wasp they help to raise) as a result of common descent; b and c are respectively the costs to the actor and benefits to the recipient.

According to this rule, under the right circumstances, it is not against Darwinian rules that some individuals direct altruistic behaviors towards their close relatives, who are more likely to share with them the same genes. Talking from a gene's eye view perspective, a gene with altruistic effects can spread in a population if it favors the fitness of those individuals who bear copies of that same gene. This means that, in order for a gene to transmit copies of itself in a population, it is not necessary that it gets transmitted directly in the offspring of the individuals bearing it. If its bearers helps members in the population that are related to itself, then it indirectly helps the transmission of copies identical to itself in other members of the population. Thus, Hamilton wrote: "The social behavior of a species evolves in such a way that in each distinct behavior-evoking situation, the individual will seem to value his neighbors' fitness against his own according to the coefficients of relationship appropriate to the situation" (Hamilton 1964b, 19).

In Part II of the 1964 paper, Hamilton presented his famous hypothesis about how inclusive fitness theory might apply to the evolutionary emergence of altruistic behaviors in social insects of the order Hymenoptera (Hamilton 1964b). Insects of this order have an unusual sex determination pattern, so called haplodiploidy, as females are diploid, say they have a double set of chromosomes, whereas males are haploid, say they have only one set of chromosomes. Haplodiploidy entails that females on average share more genes with their sisters than with their own offspring. On average, in case of single insemination and in absence of inbreeding, the amount of genes that a female shares with her own sister is ¾, whereas the amount of genes that she shares with her own offspring is ½. This is why, the haplodiploidy hypothesis is also named the '¾ relatedness hypothesis' (for instance, West-Eberhard 1975).

The haplodiploidy hypothesis suggests that, the frequent evolution of sterile workers in Hymenoptera might be the result of the unusually high relatedness of hymenoptera sisters due to male haploidy, which leads to all sperm produced by a male being identical. This means that a female may well be able to get more genes into the next generation by helping the queen reproduce, hence increasing the number of sisters she will have, rather than by having offspring of her own. Although it is still unclear how much stress Hamilton put on explanatory power of this hypothesis (Segerstrale 2013), it seemed that the theory of inclusive fitness could provide an explanation of how sterility might have evolved in Hymenoptera by focusing most importantly at the ¾ relatedness between self-sacrificing workers and the brood they attend (Hamilton 1964b).

3.3 The Puzzle

According to Hamilton, after the production of the model of inclusive fitness and the elaboration of the haplodiploidy hypothesis, an important task was finding evidence that the hypothesis would actually work in nature. This meant testing whether his haplodiploidy hypothesis actually matched the complexity of the biology of Hymenoptera. The process of gathering evidence started during Hamilton's first trip to South America in 1963–1964.

Right before leaving Europe, Hamilton got notified that the Journal of Theoretical Biology had accepted for publication the manuscript that later became *The Genetical Evolution of Social Behavior I&II*. Yet, he also got notified that he needed to make major revisions. Most importantly, he had to split the paper into two parts, one containing the mathematical derivations of the model and the other one containing the biological examples (Hamilton 1996; Segerstrale 2013). Although he was not happy about having to work more on the manuscript, he had to do so while in Brazil: "Hope of escaping from this paper was one of the reasons for coming here it seems that I did not succeed" (ZIXUN/5, Hamilton to Richards, February 19, 1964). This is why, some of the observations he made in Brazil became part of the 1964 paper. Others stayed in the Notebooks and found their place in letters that Hamilton wrote to family, colleagues and friends.

3.3.1 *Brazilian Wasps*

In his application for the Darwin Fellowship justifying the reasons why he was planning on a trip to South America, Hamilton wrote:

> My primary reason for choosing South America was the great variety of species of social wasps found there. Most of them are little known but reports indicate that social features which are of the greatest interest to my theory (i.e. somewhat contradictory to it). I want to find out something about the genetical kinship existing in colonies and swarms of these wasps by marking individuals with paint, observing their egg laying etc; also to discover whether the individuals of a colony show any discriminations based on closeness of kinship in their social behavior (Z1XJO/1/5, May, 15, 1963).[1]

This short passage makes clear that Hamilton was interested in investigating wasp societies because they were somewhat contradictory to his theory. Here, by theory, Hamilton likely meant the way his mathematical model of inclusive fitness could be used to explain the evolution of altruistic behaviors in Hymenoptera by using the haplodiploidy hypothesis. This passage also shows that Hamilton's main interest was to assess the degree of relatedness, of genetic kinship, in colonies and swarms of tropical social wasps.

In the Notebook from Brazil Hamilton reported about his first day in the lab of the famous entomologist W.E. Kerr in Rio Claro and excitedly wrote: "Spent time walking around campus looking for *Polistes* nests that are hanging everywhere." (ZIX42/1/13, Notebook 1, September 13, 1963). During his trip to South America, Hamilton collected hundreds of wasps and other insects. The ones that were not stolen from his car in Nicaragua are now stored in the British Museum of Natural History in London. He closely observed and reported behaviors and experimental observations about 8 genera of wasps, 8 genera of bees and 2 genera of ants. He collected and observed wasp nests of *Polistes fuscatus*, *Polistes canadiensis*, *Mischocyttarus cassanunga*, *Mischocyttarus dormans*, *Polistes cinerascens*, *Apoica pallida*, *Protopolybia minutissima* and others (Z1XUN/15).

3.3.2 *Polygyny and Poliandry*

Two were the main biological features of wasp social life that puzzled Hamilton and that made their social systems an anomaly for his theory: polygyny, say the presence at nest foundation of multiple egg-laying and potentially unrelated females (Hamilton 1964b), and polyandry or multiple mating, say the fact that the reproductive individuals mate multiple times (Hamilton 1964b). Both polyandry and polygyny lower the degree of relatedness in colonies. Yet, the haplodiploidy hypothesis suggested that the frequent evolution of sterile workers in Hymenoptera might be

[1] In the rest of this article, when quoting material from the W.D. Hamilton Archive in The British Library in London, I will directly use the sorting number in capital letters and numbers.

the result of the unusually high relatedness of hymenoptera sisters due to male haploidy. Therefore, both polyandry and polygyny posed important challenges to Hamilton's explanations of the evolution of social life.

Poliandry is common in Hymenoptera, and especially prevalent in wasps. In the Application to the Darwin Fellowship, after talking about his intention to work on wasps, Hamilton wrote that one of the main reasons for him to go to Brazil and spend time in the lab of the famous entomologist W.E. Kerr was: "… to learn Kerr's technique of determining the occurrence of multiple mating Hymenoptera by sperm counts, and to apply it to a variety of social and semi-social Hymenoptera. This question of multiple mating has an important bearing on my idea of how social behavior might have evolved in the group" (Z1XJO/1/5, May, 15, 1963).

In Part II of the 1964 work, Hamilton reported the problematic case of multiple mating in Hymenoptera and wrote: "Clearly multiple insemination will greatly weaken the tendency to evolve worker-like altruism" (1964b, 33). By mating with multiple males, the queen's progeny becomes very genetically diverse. Thus, multiple mating decreases the degree of relatedness among self-sacrificing workers and the queen or her brood. This makes it difficult to explain why the workers would perform altruistic behaviors in such colonies. According to Hamilton, this clearly posed a challenge to the power of the haplodiploidy hypothesis and raised the problem of how self-sacrificing behaviors in wasps were even possible, if the colonies are made out of workers with different genetic origins. He wrote: "It does seem at first rather surprising that altruism towards sisters so much less related than full sisters can be maintained at its observed pitch of perfection" (1964b, 34).

Besides multiple-mating, Polygyny, the phenomenon of multiple egg-laying queens was another aspect that made wasp societies somewhat contradictory to Hamilton's theory. Hamilton argued that: "Clearly this social mode presents a problem to our theory. Continuing cycle after cycle colonies can come into existence in which some individuals are almost unrelated to one another." (Hamilton 1964b, 36). In polygynous colonies, the workers attend a brood produced by more than one female, which means that they are not attending a brood composed only by full sisters. Also in this case, the puzzling nature of polygyny is due to the fact that it lowers the degree of relatedness in the colony. According to Hamilton, rather than favoring altruistic behavior, polygyny seems to be favorable to the spreading of genes causing selfish behaviors, which would lower the efficiency of social life. Yet, and here is the puzzle: "(…) it does not seem to do the colonies much harm and the species concerned are highly successful in many cases" (Hamilton 1964b, 36).

Hamilton looked at the polygyny puzzle in two groups of the Vespidae family: the subfamily Polybiinae and the genus *Polistes* belonging to the subfamily Polistinae (Richards and Richards 1951). Polybiinae are mostly swarm founding wasps, where colony reproduction happens by swarming with several fertilized queens. In most species of this subfamily, there are at least several queens that engage in egg-laying on each nest (Ducke 1914; Richards and Richards 1951). In the case of swarm founding wasps, the problems posed by polygyny is extremely severe due to the high number of egg laying queens. In this case, Hamilton noticed, the probability is higher to obtain colonies where individuals are almost unrelated to each other.

The situation is slightly different in the partially polygynous and partially monogynous Polistes, as wasps of this genus show different modes of nest foundation depending on the climate and on the latitude (Richards and Richards 1951). Hamilton wrote: "The geographic distribution of the association phenomenon in *Polistes* is striking" (Hamilton 1964b, 37). Polisteshave different modes of colony foundation that go from mostly monogynous in colder regions to polygynous in the tropics. In temperate regions, usually, several wasps contribute to the foundation of the nest, but one of them becomes the only egg-laying and dominant one (Pardi 1942, 1948). The rest of the wasps, the auxiliaries or subordinates cannot reproduce and, if any, they succeed in laying only a few eggs. But with many of *Polistes* species, mostly in warmer climates, nest foundation is carried out by two or more fertilized queen-sized wasps (Pardi 1942, 1948). This condition led in earlier studies on Polistes to discuss whether the polygyny at nest foundation in *Polistes* was real or fictitious (Pardi 1942).

Hamilton thought that, hierarchy formation in *Polistes* posed a challenge to his hypothesis, as he found difficult to explain the ready acceptance of non-reproductive roles by the auxiliaries (Hamilton 1964b). Although Hamilton was not able to explain this phenomenon, he mentioned that: "There is good reason to believe that the initial nest-founding company is usually composed of sisters, which brings the phenomenon closely into line with the pleometrosis of the polybiines" (Hamilton 1964b, 37). Some years later, Hamilton was still puzzled by whether or not the foundresses of Polistes nests were actually sisters or not. In a letter to M.J. West-Eberhard who had recently published an article in *Science* about the evolution of social behavior in *Polistes* (West-Eberhard 1967), Hamilton asked West-Eberhard about the evidence she had the degree of relatedness among the founding wasps in two different species of *Polistes* (West-Eberhard 1967, 1969).

3.4 Grappling with a Puzzle

Both polygyny and polyandry constitute Hamilton's touchstone puzzle. He addressed this puzzle by looking into the behavioral, biological and social mechanisms that would increase the degree of relatedness in the colony, the r in his formula. Already in the 1964 paper, Hamilton's surmised that one way to increase the degree of relatedness in the case of polygynyc associations was inbreeding and that inbreeding depended on the high viscosity of wasp population structure. He wrote: "However, it does seem necessary to invoke at least a mild inbreeding if we are to explain some of the phenomena of the social insects – and indeed of animal sociability in general – by means of this theory. The type of inbreeding which we have in mind is that which results from a high viscosity of population or from its actual subdivision into small quasi-endogamous groups" (Hamilton 1964b, 65).

Although puzzling, social wasps offered Hamilton unique opportunities for observation and experimentation. "There is no doubt that in [it is] a very unusual biological situation and with a species which is highly social in a rather flexible and human way, it offers great possibilities for observation and experiment." (ZIX42/1/13, Notebook 1, November 22, 1963). This is why, in his first trip to Brazil, Hamilton engaged in numerous observations and some attempts to experiment on nests of both Polistinae and Polibinae, with the hope to test whether his ideas about inbreeding and viscosity could actually work.

3.4.1 Observations and Attempts to Experimentation

In Brazil, Hamilton wanted to figure out how viscosity could lead to inbreeding, and in this way, contribute to make the degree of relatedness higher. Before the discovery of genetic markers and techniques to assess the genetic relationship of individuals in a colony, finding out about kinship and relatedness was not an easy task. Still, Hamilton tried to find clues that would hint at the level of relatedness in the colonies and how that would affect the behaviors of the individuals.

In some homing experiments, Hamilton looked for a correlation between relatedness and the distribution of the population in a certain area. In this way, Hamilton wanted to see whether average relatedness could correlate with geographic proximity (ZIX55/1/3, Notebook 1). He looked into the viscosity of wasp populations trying to figure out whether or not the offspring would disperse slowly from their site of origin or if they would tend to stay close to the nest. In some of these experiments he tried to test the flight range of the wasps so as to see if there were differences between the capacity to fly far away and the level of altruistic behaviors in the colonies (ZIX55/1/3, Notebook 1). Also, Hamilton performed some transference experiments where he would introduce wasps unrelated to the rest of the colony (or related as a control) to see the different reactions. In these experiments he tried to find out if wasps from distant localities are less likely to be accepted on a nest than wasps from nearby nests (ZIX55/1/3, Notebook 2).

Beside transference and homing experiments, due to his interest in viscosity and population structure, Hamilton started looking into the relationship between nest architecture and social habits (Z1X83/1/10, Hamilton to West-Eberhard, October 5, 1967). This is why he collected, took pictures and produced sketches of hundreds of wasp nests while in Brazil. Hamilton collected only specimens close to the nest and always specified where he found it and what it was doing in relation to the other wasps so as to get a better idea of how the social habits of each colony were both reflected in and influenced by the architectural structures of the nests. Hamilton got very interested in nest architecture and was planning on writing a substantial article on the topic, but unfortunately he never made it.

But Hamilton did not limit his observations and experiments to behavioral observations. Right upon arrival in Rio Claro in Kerr's lab, Hamilton learned new techniques about how to carefully dissect wasps in order to find out about the physiological status of their internal organs. On the third day he was in Rio Claro, he performed his first wasp dissections and started a very accurate Index Card system to collect physiological observations about the wasps he had collected (Z1XUN/15). Each card was organized in five columns containing information about specific aspect of the biology of that wasp.

The first column was a simple number identifying the wasp. Hamilton would use this numbers in his field notes when referring to those wasps, creating in this way a complex reference system between the notes in his books and the index cards. The second category was about whether the wasp collected was a male or a female. In the case of female wasps, he always pointed out whether the ovaries were developed or underdeveloped as a sign for the possible social status of each wasp. The third column was about the status of the spermatheca and whether or not it was packed with sperm. This column was essential to Hamilton's interest in polyandry. This is why he learned specifically how to dissect the spermatheca from Kerr and his students. The status of development of the fat bodies, whether they were well developed or not occupied ususally the fourth column. Previous studies in wasp physiology had described the correlation between development of fat bodies, ovarian development and social status of each wasp. Finally, the last column described wing length and length of the first lergite as well as the presence and status of the hamuli on wings. This last information helped Hamilton to figure out the relationship between flight range and geographic distribution of the wasps in a given area.

3.4.2 Doubts and Adjustments to the Model

Although he collected a great amount of data and observations during his first trip to Brazil, Hamilton admitted that he should perform those again more systematically: "There are several other experiments that I would like to make to amplify and confirm the results I obtained last time. I would like to make more careful series of transferance experiments with adequate controls to find out if wasps from distant localities are less likely to be accepted on a nest than wasps from nearby nests. I also want to make some homing experiments to find out the flight range." (Z1X83/1/10, Hamilton to West-Eberhard, October 5, 1967). Hamilton was planning on doing so in his second trip to Brazil in 1968–1968, but also the results of this second trip did not turn out to provide conclusive answers with respect to the wasp puzzle.

Still in the late 1960s and early 1970s, Hamilton was not totally satisfied about how his theory could match the complexity of the biological world. In a letter again to M.J. West-Eberhard Hamilton confessed his frustration: "I am still quiet unhappy about that theory in general, feeling that although it seems to be on the right lines it needs considerable elaboration before it begins to match the complexity of biological situations" (Z1X83/1/10, Hamilton to West-Eberhard, October 5, 1967). In a slightly different tone, in a letter from 1966 to W.E. Kerr, he wrote: "I am still cautios about the value of my theory in the case of the social Hymenoptera more cautios, I think, than Prof. E.O. Wilson who so generously supported it at a recent meeting of the Royal Entomological Society" (ZIX89/1/1, Hamilton to Kerr, May 6, 1966).

In the same letter to West-Eberhard, Hamilton expressed his ideas about inbreeding and its relation to viscosity: "I am now inclined to place relatively more weight on inbreeding as a factor raising the coefficient of relationship and so facilitating social evolution and less on the special features of male haploid relatedness than I was when I wrote those papers" (Z1X83/1/10, Hamilton to West-Eberhard, October 5, 1967). In the late 1960s, after his second trip to Brazil, Hamilton was asked to work on a paper where he could explain and revise his 1964 arguments and ideas about "how relatedness affects the evolution of social insects" (Hamilton 1996, 255). Beside other modifications, in the 1972 paper *Altruism and Related Phenomena, mainly in Social Insects*, Hamilton included inbreeding into his formulation of the mathematical model of inclusive fitness and newly discussed the situation of Polygynous wasp colonies (Hamilton, 1972).

Hamilton was able to incorporate inbreeding into the model thanks to the re-derivation of his formula developed during his collaboration with George Price (Hamilton 1996, 256). Although the model was improved, that did not actually mean that it was able to address the puzzle posed, mostly by Polygyny, to the haplodiploidy hypothesis. Inbreeding might be a good way to explain the puzzle as it raises the degree of relatedness in the colony. Hamilton argued: "Unless there is a very high degree of inbreeding, why does not intracolony selection for queen-like behavior break down the system? Why do workers work so willingly and by what device are the fierce struggles for dominance that occur, for example, in queenless *Apis* and *Vespa* colonies prevented?" But Hamilton admitted as well that: "If inbreeding is the answer, would we not expect more genetic diversity between colonies, relative to uniformity within colonies, than we actually observe?" and added "But these questions cannot be answered yet." (Hamilton 1972, 216).

So, adding inbreeding to the model, did not actually help to provide a solution of the puzzle. In fact, Hamilton in the 1972 review would still claim that: "In my opinion, the polygyny in Polybiini, [...] provides the most testing difficulty for the interpretation of the social insect pattern which is offered in this review" (Hamilton 1972, 216). Already in the letter to Kerr mentioned above, Hamilton admitted that looking into haplodiploidy was likely not enough to explain the evolution of biological altruism: "I feel that the male-haploidy cannot be more than half the story, and that the other half must involve the classical concepts of the fabricating, provisioning, long-lived Hymenoptera." (ZIX89/1/1, Hamilton to Kerr, May 6, 1966). This means that Hamilton was thinking about including other factors and processes in the explanation of how altruistic behaviors might have emerged in evolutionary history. Still, such factors would have had to contribute, according to Hamilton, to raise the relatedness in the colony (Hamilton 1972).

3.5 Conclusions

This article has provided an overview on Hamilton's attempts to deal with the puzzling social behaviors of wasp societies. In the process of finding evidence for his hypothesis of haplodiploidy, Hamilton dealt with this puzzle rather unsystematically. Still, he had an idea. He surmised that there was a correlation between viscosity and inbreeding. He thought that, due to the high viscosity observed in wasp colonies, increases in inbreeding could affect the overall relatedness in the population. He tried to find correlations between viscosity and inbreeding and addressed the puzzle by performing experiments, conducting observations and dissecting the wasps. Eventually, Hamilton updated his mathematical theory of inclusive fitness an included inbreeding. Nonetheless, the wasp puzzle remained and Hamilton was not actually able to solve it.

Although Hamilton's investigations of wasp colonies focused mostly on finding out about the relationship between viscosity and inbreeding, he actually started producing evidence on many factors that could influence the degree of relatedness in the colony, from flight range to possible number of queen matings and dispersal capabilities of different wasps. Trying to find evidence in favor of his theory and of his hypothesis, Hamilton started exploring the biological mechanisms underpinning the coefficient of relatedness and tried to bind and connect those phenomena within the framework of his theory. Consequently, still with an open mind, he eventually modified the theory in the light of the evidence gathered during these explorations. In this way, Hamilton suggested an open approach for the understanding of social life and its evolution that would inspire and guide future generations of evolutionary biologists and scholars of wasps.

Acknowledgements I am greatly thankful to J.B. Grodwohl for comments on previous drafts of this article and to Jonathan Pledge, from W.D. Hamilton's Archive at The British Library in London, for his invaluable help. I also thank all the participants of the Integrated HPS Conference in Vienna (July 2014). Finally, I wish to thank the Center for Biology and Society, the School of Life Sciences and the Graduate and Professional Student Association at Arizona State University for financially supporting two trips to the Hamilton's Archive in 2014.

Bibliography

Charnov, E.L. 1977. An Elementary Treatment of the Genetical Theory of Kin-Selection. *Journal of Theoretical Biology* 66 (3): 541–550.

Ducke, A. 1914. Uber Phylogenie und Klassifikation der sozialen Vespiden. *Zoologische Jahrbücher, Jena: Abteilungen für Systematik, Ökologie und Geographie der Tiere* 36: 303–330.

Hamilton, W.D. 1963. The Evolution of Altruistic Behavior. *American Naturalist*: 354–356.

———. 1964a. The Genetical Evolution of Social Behaviour. I. *Journal of Theoretical Biology* 7: 1–16.

———. 1964b. The Genetical Evolution of Social Behaviour. II. *Journal of Theoretical Biology* 7: 17–52.
———. 1972. Altruism and related phenomena, mainly in social insects. *Annual Review of Ecology and Systematics* 3: 193–232.
———. 1996. Foreword to "Natural History and Evolution of Paper-Wasps", ed. S. Turillazzi and M.J. West-Eberhard, v–vi.
Pardi, Leo. 1942. Ricerche sui Polistini V. La poliginia iniziale di Polistes gallicus (L.). *Bollettino Istituto di Entomologia Università di Bologna* 14: 1–106.
Pardi, L. 1948. Dominance Order in Polistes Wasps. *Physiological Zoology*: 1–13.
Richards, O.W., and M.J. Richards. 1951. Observations on the Social Wasps of South America (Hymenoptera Vespidae). *Transactions of the Royal Entomological Society of London* 102 (1): 1–169.
Segerstrale, U. 2013. *Nature's Oracle: The Life and Work of WD Hamilton*. Oxford University Press.
West, M.J. 1967. Foundress Associations in Polistine Wasps: Dominance Hierarchies and the Evolution of Social Behavior. *Science* 157 (3796): 1584–1585.
West-Eberhard, M.J. 1969. The Social Biology of Polistine Wasps. *University of Michigan, Museum of Zoology. Miscellaneous Publications* 140: 1–101.
———. 1975. The Evolution of Social Behavior By Kin Selection. *Quarterly Review of Biology*: 1–33.

Chapter 4
The First Century of Cell Theory: From Structural Units to Complex Living Systems

A Look from 1940 at 100 Years of Cell-Theory

Jane Maienschein

In his introduction to the volume entitled *The Cell and Protoplasm* in 1940, the editor Forest Ray Moulton noted that the American Association for the Advancement of Science was publishing the volume as the product of a symposium, held in 1939, to celebrate the centennial of Matthias Schleiden and Theodor Schwann's 1838 cell theory.[1] Because of the rich history of thinking about cells up to that time, "In a sense the Cell Theory is not new." Yet, Moulton suggested, "In another sense the Cell Theory is always new, for every discovery respecting this primary and essential unit of living organisms, both plant and animal, has raised more questions than it has answered and has always widened the fields of inquiry."[2] The volume set out to show both what was already well-established and what was new.

By 1940, discussion of cells involved looking at a predictable list of topics, including the way cell walls delineate individual cells, contents of cells including nucleus and cytoplasm and organelles, and environment interactions both internal to and external to each cell. With those came considerations of biochemistry and cell physiology. Less predictable are the chapters on microbiology, viruses, enzymes, hormones, and vitamins. The choice of topics and of contributors makes clear just how many questions remained in the middle of the twentieth century about cells and especially how they interact within organisms.

Cell structure and function clearly affect the organism, but just what the causal connections were remained unclear. By mid-century, two different views co-existed. As Lester Sharp put it in his 1943 textbook *Fundamentals of Cytology*, the first held that cells are the agents of organization of the organism, while the second or

[1] Schleiden (1838); Schwann (1839).

[2] Forest Ray Moulton (1940), Foreword.

J. Maienschein (✉)
School of Life Sciences, Arizona State University, Tempe, Arizona 874501, USA

Marine Biological Laboratory, Woods Hole, MA, USA
e-mail: maienschein@asu.edu

F. Stadler (ed.), *Integrated History and Philosophy of Science*, Vienna Circle Institute Yearbook 20, DOI 10.1007/978-3-319-53258-5_4

organismal view placed the agency with the organism as a whole and emphasized, "the primacy of the whole, cells when present being important but subsidiary parts."[3] In both cases, evolutionary history was thought to have played important roles in shaping the patterns that emerged. But do the cells drive the organism, or does the organism drive the cells? What research and what epistemological assumptions had led to this mid-twentieth century question?

4.1 The Standard Story

The familiar story, recounted in textbooks and cell biology courses, declares that Schleiden and Schwann invented the cell theory. Standard accounts tell of these two German innovators, one working on plants and the other on animals, coming up with *the* theory that cells are *the* fundamental unit of life. In 1838, the story goes, they put together the available evidence and reasoning to develop what they called the *Zellentheorie* to ground all of biology, and they were the first to do so.

Everybody likes a good myth, and this one does its job. Schleiden and Schwann did, in fact, respectively study plants and animals and did see and describe cells. They were not the first, however, but drew on earlier observations by Robert Hooke, Anthony Leeuwenhoek, and many others to establish the idea of structural cellular units as bounded by walls. Internally, these seventeenth century microscopists held that cells might consist of some fluid-like or gel-like substance or they might be vesicles full of nothing more than air. What was important is that they each observed vesicles with walls and structure, and came to call them cells.[4]

For decades, textbooks have referred to these two as the fathers or founders of the cell theory, as if they had articulated a theory of the basic units of life. In fact, Schleiden and Schwann saw *basic units of living organisms* but not *basic living units*. That is, they did not clearly consider those cells to be "alive" since they did not reproduce themselves but were thought to arise at least at times through a sort of crystallization. The standard story misses the distinction and ascribes more agency and properties of life to the cell then Schleiden and Schwann did themselves. The myth has not changed much, despite new historical studies after a century of cell biology.

4.2 Writing a New History Around 1950

In the mid twentieth century, several researchers set out to look more closely at the historical record by studying what the cell theory meant and what each contributor had actually done. In 1948, Oxford University cytologist John Randal Baker began a series of five essays in the *Quarterly Journal of Microscopical Science*. Under the

[3] Sharp (1943), p. 21.

[4] Harris (1999). Chapters 1 and 2 on early microscopists and early theories.

title "The Cell-theory: a Restatement, History, and Critique," Baker looked closely at the primary sources. He began with the point that "Several zoological text-books published during the last two decades have cast doubts on the validity of the cell-theory." Baker resolved to review the attacks, the nature of the evidence, and to establish the current status of the cell theory. He found different critiques attacking different aspects of what was lumped together as the cell theory, and he found some of the attacks justified while others were not.[5]

Baker broke down the larger theory into what he called "propositions," focused on the shape, characteristics, origin, development, and individuality of cells, and claims about the relationships of multiple cells in multicellular organisms. Baker carried out a tremendous service in clarifying what was at issue with cell biology. He showed that Schleiden and Schwann each, in different ways, made assumptions about how cells originate and/or about their structure and nature that went beyond their data. In some cases, they worked with inadequate microscopic tools; in other cases, they started with strong assumptions about what they should see and then somehow persuaded themselves that they actually saw what they wanted – whether it was really there or not. Thus, Schleiden was confident that he actually observed cells crystallizing around a nucleus, even when further investigation using contemporary tools show that that is not true.

Baker's scholarly essays appeared from 1949 to 1955 and showed who had said and thought what, when, and why. Shortly after, in 1959, the Cambridge University anatomist Arthur Hughes added his *A History of Cytology*. Like Baker, Hughes sought to clarify the development of understanding of cells. Hughes emphasized cytological methods of investigation alongside the theories, with special emphasis on the nucleus and cytoplasm.

In his *The Birth of the* Cell in 1999, Henry Harris went over much of the same ground as Baker and Hughes, but with considerably more subtlety in scholarship and interpretation. He re-read the original sources, and furthermore had the benefit of an additional half century of biological discovery and reflection. Harris pointed to an 1843 quotation from the French microscopist François-Vincent Raspail to show that Schleiden and Schwann did not have the only word even at their time. "Give me an organic vesicle endowed with life," Raspail said, "and I will give you back the whole of the organized world."[6] For Schleiden and Schwann, the cells were units of structure, while for Raspail they were units of life. Claiming that the German story focused on cell structure had come to dominate, Harris called for recognizing Raspail's vision of the cell as a "kind of laboratory" which allowed life to develop and reside within the cell.

Yet despite Raspail's ideas and Harris's efforts to revive them, the German interpretation has continued to dominate cell biology and our historical reflections about it. It is therefore worth revisiting again some key historical contributions, and to try to get at the distinction of cells as units of living systems or cells as themselves living systems.

---

[5] Baker (1948), page 103.

[6] Raspail (1833), Quoted in Harris (1999), p. 32.

4.3 Putting the Life in Cells

Several main themes in the nineteenth and into the twentieth century help illuminate the difference understandings. First is understanding of Schleiden and Schwann's cell theory, then fertilization, the relative roles of nucleus and cytoplasm, cell lineage and development, cell-cell interactions, regeneration, and cell culture outside of the organism. Here, we can look briefly at each of these points.

4.3.1 Cell Theory

Schleiden and Schwann in 1838 gave cell theory a name and declared first that cells exist and are constituents of living organisms, and second that the theory might help explain individuality of organisms as clusters of connected cellular units. As Henry Harris aptly puts it, Schleiden's long article on cells in plants "does not make pleasant reading." Fortunately, Harris helps us digest the key points. Schleiden, like Schwann, saw the nucleus (which he called the cytoblast) as centrally important. They each held that the nucleus is the structure that appears first and then generates the cell. From the moment he had encountered Schleiden's ideas during a conversation at dinner one evening, Schwann claimed, "I devoted all my energies to demonstrating the pre-existence of nuclei in the formation of cells."[7] Sometimes the nucleus exists alone and the cell crystallizes around it, sometimes the cells divide and each has a nucleus.

Since this is a central point in the reasoning, it is worth quoting Schleiden at some length, and Harris provides an excellent translation explaining how cells arise:

> As soon as the cytoblasts have attained their full size, a delicate, transparent vesicle is formed on their surface. This is the young cell which to begin with appears as a very flat segment of a sphere, with its planar side constituted of the cytoblast and its convex side by the young cell which is superimposed on it much like a watch glass on a watch. … Little by little the whole cell now grows out over the edge of the cytoblast and soon becomes so big that the latter eventually appears as no more than a small body enclosed within one of the parietal walls.

Schwann accepted Schleiden's interpretation, even though he saw cell division in addition to what he interpreted as crystallization in the animal cells he studied. Therefore, both emphasized the role of the nucleus as a "cytoblast," or literally cell-developer. Each individual cell emerged and cells then served as structural parts of the organisms in which they resided. And yet, Harris emphatically notes, "I think it is fair to say that no part of the scheme proposed by Schleiden turned out to be correct."[8] Harris provides suggestions about why these German cytologists gained so much attention at the time and later, but for our purposes, the point is that they

[7] Harris (1999), p. 96.

[8] Harris (1999), p. 98. Translated from Schleiden (1838).

emphasized the structure of the cells and saw them as not the only, but some of the structural parts of the living organisms. In fact, much of what has been credited to Schleiden and Schwann came later.

The decades following brought a great deal of additional observation as well as interpretation. Those years also brought improvements in both microscopes and microscopic techniques, as Hughes discusses in detail. For studying cells, it makes a big difference what one can see and how well one can see it; and making sure that others can see the same thing is especially important. Better lenses reduced chromatic aberrations, and better fixing, staining, and slicing methods improved standardization of specimens to improve consistency of observation. But it wasn't just being able to see more that mattered. It was also looking more carefully, and with a more open mind than Schleiden and Schwann seem to have had.

4.3.2 Fertilization

Aristotle thought that fluids from the male and female come together and somehow combine, so that form and function emerge only gradually in a way he called epigenetic. The idea of a process of fertilization required first the idea of an egg, so that there was something to be fertilized. Karl Ernst von Baer played an important role here when he observed a mammalian egg for the first time. Chick and frog eggs were big and obvious, but it wasn't clear whether all organisms had eggs or not. Von Baer thought they must and went looking, offering the first clear description in 1827 with an egg from a dog.[9] Animals start from eggs, it seemed.

Furthermore, those eggs seemed to be fertilized by spermatozoa, yet it took a number of people and many observations to observe that a sperm cell actually combines with an egg cell. George Newport wrote three lengthy descriptions of his observations and experiments to discover how the spermatozoa "impregnate" eggs, concluding that they carry some force of "vitalization," or process of coming alive.[10] It took a few more decades for Oscar Hertwig to report observations of sperm cells actually entering into, combining with, and thereby fertilizing egg cells. He observed in detail each step of the entry process, as well as appearance of two nuclei and then reduction and division into one nucleus for the fertilized egg.[11] By the time of Hertwig's work in the 1870s, it had become clear that fertilization involves the process of two cells coming together to make one cell.

Edmund Beecher's *Atlas of Fertilization and Karyokinesis*[12] in 1895 presented the process of fertilization photographically. He showed the details of sea urchin egg cells combining with sperm cells, reduction division of nuclei, chromosomal and cytoplasmic changes in preparation for cell division, and then the process of

[9] von Baer (1827).

[10] Newport (1851/1853/1854).

[11] For discussion of this point, see for example: Churchill (1970).

[12] Wilson (1895).

cleavage itself. One cell divides into two, into four, and so on. Wilson gave his reader photographs, taken in collaboration with the photographer Edward Leaming, and sketches of the key details to highlight essential features of the process. By the end of the nineteenth century, it was clear to biologists that an individual organism's life began as cells, which underwent fertilization and then divided and differentiated into a complex organism. The egg and sperm cell, and the cells resulting from cell division had begun to have a biological life of their own.

4.3.3 Cells from Other Cells

The close detailed observations of fertilization and the cell division that followed raised questions about the nature of the living organism. If cells crystallize around nuclei and out of surrounding material, as Schleiden and Schwann had said, then the life of the organism has to come from somewhere. Perhaps it is preformed in the nucleus somehow, or is spontaneously generated, or arises gradually due to some vital force that we have not observed directly. In contrast, if each cell comes only from other cells, then the "life" and the beginnings of the form and function of an individual are in some sense already there from the beginning with that first cell that came from a previously living organism.

Robert Remak rejected Schleiden and Schwann's idea that cells behave like crystals, insisting that they are completely different.[13] The egg is a cell and is the starting point for development of each organism, Remak insisted. Furthermore, that initial cell goes through division to produce more and more cells, which then work together to make up the organism. Finally, since the egg itself came from the previous generation and is alive in the sense of having the capacity to become fertilized and to divide, the living cell comes from another living cell. Life comes from life, and never from some intracellular and inert material.

Rudolf Virchow went further, and Harris discusses the relationship between Virchow and Remak. The two began as friends, but on just this point about the origin of cells and therefore of life, the two diverged vehemently with Remak eventually accusing Virchow (with considerable evidence on his side) of plagiarism.[14] In his work on *Cellular Pathology*, Virchow nonetheless showed decisively through empirical observation that cells divide and give rise to other cells. In his book, quickly translated into English and widely read, Virchow famously declared that "omnis cellula e cellula" and thereby that life comes from other life.[15] Cells had taken on a different status than they had for Schleiden and Schwann: they were not just structural units of living organisms; now they were living in themselves and the living units that make up organisms.

---

[13] Robert (1855).

[14] Harris (1999), p. 132.

[15] Virchow (1858). Translated as: *Cellular pathology*, London: John Churchill, 1859.

4.3.4 *Cytoplasm and Nucleus*

By the end of the nineteenth century, Theodor Boveri, Oscar Hertwig, and Edmund Beecher Wilson, among many others, were studying cells inside and out. It was clear that the cell has structure, a distinct bounded nucleus, liquid or gel-like cytoplasm, and other internal structures including mitochrondria and Golgi bodies, with spindle fibers, asters, and centrosomes playing important roles during cell division. Each cell has a life of its own, and these researchers were showing the ways in which it functioned as a complex dynamic system in itself and in interaction with other cells. The improved microscopic methods and careful observations in the final quarter of the nineteenth century had established this view of the cell.

Understanding the cell's role as a living system involved sorting out what was going on with heredity and development. In his work of 1893 and 1898, Hertwig solidified accumulating evidence and reasoning about the nature of fertilization, observing the details of the way in which the nucleus of the egg and sperm come together to make a new nucleus for the zygote.[16] On the first page of his volume pulling together studies of cells and tissues, he saw the cell theory as an understanding of cells as the "vital elementary units." His volume therefore provided a comprehensive theoretical and empirical framework for understanding cells as living systems. Hertwig paid close attention to the interaction of nucleus and cytoplasm in particular to get at the way that cells are alive.

Edmund Beecher Wilson's masterful 1896 study of *The Cell in Development and Inheritance* appeared about the same time as Hertwig's. For Wilson also, the cell clearly plays a foundational role for life and therefore necessarily for biology. Dedicated to Theodor Boveri, *The Cell* opened by pointing to Schleiden and Schwann and noting that "it has become ever more clearly apparent that the key to all ultimate biological problems must, in the last analysis, be sought in the cell." Furthermore, "No other biological generalization, save only the theory of organic evolution, has brought so many apparently diverse phenomena under a common point of view or has accomplished more for the unification of knowledge."[17]

By his third and final edition in 1924, Wilson acknowledged that a great deal had changed, the volume had grown from 371 to 1232 pages, and had undergone reconceptualization while seeking to retain its synthetic approach. He opened with a slightly different tone: "Among the milestones of modern scientific progress the cell-theory of Schleiden and Schwann, enunciated in 1838–39, stands forth as one of the commanding landmarks of the nineteenth century." Yet their ideas were just a "rude sketch" that led to "opening a new point of view for the study of living organisms, and revealing the outlines of a fundamental common plan of organization that underlies their endless external diversity."[18]

---

[16] Hertwig (1893/1898).

[17] Wilson (1896), page 1.

[18] Wilson (1925), page 1.

In this third edition, Wilson points to three periods since the inception of the idea of cells: focused on the basic ideas about cells and their roles, looking at development and cell division, and then bringing in the chromosome theory of heredity that introduced explanations of the causes of cell division. Wilson pointed to recent successes: "If we are confronted," he wrote in the final paragraph, "still with a formidable array of problems not yet solved, we may take courage from the certainty that we shall solve a great number of them in the future, as so many have been in the past."[19]

Together Hertwig and Wilson called attention to the cell and especially to its complexity. They helped to bring attention to, and to stimulate additional interest in, interpreting how cells function as fundamental living units. How does each cell grow, divide, differentiation, and otherwise change over time in ways that add up to a complex organized organism?

Theodor Boveri provided some answers, looking closely at the contributions of the nucleus. In 1902, for example, Boveri demonstrated that chromosomes are defined structures and furthermore that they retain their individuality through cell divisions. They divide, so that each of the daughter cells will have its own set of chromosomes after divisions, but they retain their individuality nonetheless. Observing and carefully describing the details, Boveri added immeasurably to understanding of cell division with his experimental work. He carefully controlled conditions so that he could determine what role the cytoplasm could play on its own and what role the nucleus played. He made clear that a cell without a nucleus is not a living cell, nor is a nucleus alone capable of division and differentiation.[20] The cell as a whole is a complex living system.

4.3.5 Cell Lineage

Cells each have a cytoplasm and nucleus, but they do not all look or work in exactly the same ways. To get at how differences arise and what they mean, Wilson and others carried out detailed studies of cell lineage, that is study of the lineages of cells, how one cell divided into two and so on. Wilson, Edward Grant Conklin, and others at the Marine Biological Laboratory in Woods Hole, Massachusetts, spent considerable energy collecting specimens from different species, taking them into the lab and observing every stage as each cell divided, one by one.

This work involved observing living cells, and also preserving, fixing, staining, and in short killing them in order to observe what was going on inside. They could see the changes in shape and structure, and they could see the way the chromosomes and other parts behaved during cell division and during differentiation. They could see that it depends on where each cell is located within the organism what shape it takes and how it divides in the next step. But they could only observe for so long in

[19] Wilson (1925), p. 1118.

[20] For an excellent discussion of Boveri's work, see: Laubichler and Davidson (2008).

the organism's developmental process; eventually there were too many cells, and the complexity of the whole organism made it impossible with this method to continue observing the individual cellular parts.

4.3.6 Cell-Cell Interactions

If the individual cells are the fundamental living units, and cell lineage shows that their behavior depends at least in part on their position within the developing embryo, the next question is how they make up a complex multicellular system. Getting at that required studying cell-cell interactions, and doing so required a number of epistemological assumptions about what one is actually seeing. It is challenging to observe processes that occur over time, especially when the methods for observation involve watching sequences of killed and prepared materials. Early studies of transplantation helped illuminate these processes. Cells and tissues taken from their normal placement and role in developing organisms and moved to another place, or to another organism altogether, suggested ways cells communicate with each other.

Clusters of cells that make up the eye vesicle, as Hans Spemann showed for example, could be removed from one embryo and would result in a missing eye. Or they could be transplanted to another part of the organism or to another organism and produce an eye where there would not normally have been one.[21] This work suggested that the individual cells acquire some definition or differentiation themselves fairly early on, and that they also respond to changing conditions both inside the cell and in interactions with other cells. Only gradually in the course of the twentieth and into the twenty-first century have researchers begun to understand the vast range of cell-cell interactions through chemical and hormonal signals, neural signaling across synapses, and diverse messenger systems involved in bringing together the complex cells into a complex organismal whole.

4.3.7 Regeneration

Thomas Hunt Morgan laid out the foundations for modern study of regeneration with his book of that title in 1901.[22] Regeneration provides an excellent source of, in effect, natural experimental material. Observing cell lineage could show what happens in normal cases, but much of the process remained invisible. Experimental approaches such as transplantation could yield additional information, but also had limits. Studying regeneration could reveal cases in which cells change from normal conditions. What makes it possible for a planarian to regenerate a new head or tail,

[21] Hamburger (1988).

[22] Morgan (1901).

for example? Morgan asked whether existing cells change, that is whether they somehow became re-differentiated into a different kind of cell? Or did they instead generate new cells of the right type to make heads or tails? This raised questions about whether it was something in the organism as a whole that drove the changes, or whether the cells themselves were doing the changing? What was driving the organization of the organism – the cells or the whole of interacting cellular parts? Morgan captured these questions and understood that getting at what causes regeneration to occur in some animals for some conditions could reveal a great deal about development and about the role of individual cells.

4.3.8 Cell Culture

Ross Granville Harrison removed cells from frogs and transplanted them not to another part of that frog or even to another frog, but rather into a culture dish. The neuroblast cells he transplanted then differentiated as nerve fibers, which he interpreted as developing in just the same way they would have normally within the organism.

This experimental approach allowed a test of how cells develop on their own in an artificial culture medium but not as part of the organism. This kind of tissue and cell culture seemed to answer questions about the extent to which cells are alive and self-differentiating, in contrast to parts of organisms that determine their reactions. Cells must be self-determining to a very large extent, guided by some internal factors and responding to environmental cues. In turn, this conclusion raised questions about how the self-organizing cells then connect physically or communicate biochemically or in some other way with other cells: how do the parts make up the whole interactive and dynamic organism? How does the cellular system work in connection with the organismal system?

One idea held that some sort of protoplasm lies outside the cells and connects them. In the context of his challenges to the cell theory as sufficient explanation for organization in life, Adam Sedgwick was still invoking this idea through the end of the nineteenth century, as Baker discusses.[23] The idea of a reticulum or syncytial connections proved attractive, because it offered an explanation for how cellular parts work together as an whole system. Physical connections could make the parts into a network. This reasoning held for the nervous system in particular. At the end of the nineteenth century, researchers argued about whether the nervous system is there from the very beginning in a sort of reticulum that then grows larger while maintaining its structure. In contract, the neuronal theory held that individual neuroblast cells develop nerve fibers that grow out and make connections over time. They only gradually make up the nervous system.

Harrison accepted the second, neuronal, theory. His culture experiments described above showed the ways that individual neuroblast cells grow out and

---

[23] Baker (1948), p. 175. Sedgwick (1894).

make connections. In fact, his work was taken by many as having resolved the question in favor of the action of individual cells working together. Yet a few such Camillo Golgi never gave up their convictions that the system was inextricably interconnected from the beginning. He simply could not see how to explain the complexity of the nervous system otherwise.

The discussions were part of persistent debates about whether development is more preformationist, that is laid out from the very beginning in a preformed way, or epigenetic, that is arising only gradually over time.[24] The epigenetic view requires an explanation for how the individual cells arise and how they make up a whole organism. Where does the organization and where does the life come from if the separate and individual cells come together to make a whole?

4.4 Conclusion

Cells started conceptually as basic structural units of living organisms, arising through crystallization from non-living matter. By the end of the nineteenth century, they had acquired a life of their own and were seen as a complex living system in themselves. As we saw in the reflections after a century of the cell theory, questions persist about the extent to which and ways in which the cells organize themselves and add up into complex organisms, rather than having the organism as a whole organize the cells. Recent research with stem cells and induced pluripotency have complicated the questions still further, suggesting that cells have a tremendous capacity to respond to changing environmental conditions. As Moulton said in 1940 as editor, despite our advances in cell biology, understanding the cell continues to provide us with new insights and to challenge existing assumptions.

Acknowledgements Thanks to the &HPS organizing committee and to Friedrich Stadler to present this paper in 2014 in Vienna. And thanks to the National Science Foundation for support through a series of grants funding background research. I also appreciate my collaborators Manfred Laubichler, William Aird, and Karl Matlin for working together on understanding the history of cell biology.

References

Baker, John Randal. 1948. The Cell-Theory: A Restatement, History, and Critique. Part I. *Quarterly Journal of Microscopical Science* 80: 103–125.

Churchill, Frederick B. 1970. Hertwig, Weismann, and the Meaning of Reduction Division Circa 1890. *Isis; An International Review Devoted to the History of Science and Its Cultural Influences* 61: 429–457.

Hamburger, Viktor. 1988. *The Heritage of Experimental Embryology: Hans Spemann and the Organizer*. New York: Oxford University Press.

[24] For more discussion of this and related topics, see Maienschein (2014).

Harris, Henry. 1999. *The Birth of the Cell*. New Haven: Yale University Press.
Hertwig, Oscar. 1893. *Die zelle und die gewebe. Grundzüge der allgemeinen anatomic und physiologie*. Volume 1 (1893), volume 2 (1898). Jena: G. Fischer.
Laubichler, Manfred D., and Eric H. Davidson. 2008. Boveri's Long Experiment: Sea Urchin Merogones and the Establishment of the Role of Nuclear Chromosomees in Development. *Developmental Biology* 314: 1–11.
Maienschein, Jane. 2014. *Embryos Under the Microscope. Diverging Meanings of Life*. Cambridge: Harvard University Press.
Morgan, Thomas Hunt. 1901. *Regeneration*. New York: Macmillan.
Moulton, Forest Ray, ed. 1940. *The Cell and Protoplasm*. Washington, DC: American Association for the Advancement of Science, The Science Press.
Newport, George. On the Impregnation of the Ovum in the Amphibia. *Philosophical Philosophical Transactions of the Royal Society of London*, 3 Series: (1851) 141: 169–290; (1853) 143: 233–290; (1854) 144: 229–244.
Raspail, Françoise-Vincent. 1833. *Nouveau système de chimie organique, fondé sure des methods nouvelles d'observation*. Paris: Ballière.
Robert, Remak. 1855. *Untersuchungen über die Entwicklung der Wirbeltiere*. Berlin: G. Reimer.
Schleiden, Matthias. 1838. Beiträge zur Phytogenesis. *Müller's Archiv für Anatomie, Physiologie und wissenschaftliche Medicin*: 137–176.
Schwann, Theodor. 1839. *Mikroskopische Untersuchungen über die Übereinstimmung in der Struktur und dem Wachsthum der Thiere und Pflanzen*. Berlin: Reimer.
Sedgwick, Adam. 1894. On the Inadequacy of the Cellular Theory of Development. *Quarterly Journal of Microscopical Science* 37: 87–101.
Sharp, Lester W. 1943. *Fundamentals of Cytology*. New York: McGraw-Hill.
Virchow, Rudolf. 1858. *Die Cellularpathologie in ihrer begründung auf physiologische und pathologische gewebelehre*. Berlin: A. Hirschwald.
von Baer, Karl Ernst. 1827. *De Ovi Mammalium et Homini Genesi*. Lipsiae: Leopoldi Vossii.
Wilson, Edmund Beecher. 1895. *An Atlas of the Fertilization and Karyokinesis of the Ovum*. New York: Macmillan.
———. 1896. *The Cell in Development and Inheritance*. New York: Macmillan and Company.
———. 1925. *The Cell in Development and Heredity*. New York: Macmillan.

Chapter 5
Theorizing the Distinction Between Solids, Liquids and Air: Pressure from Stevin to Pascal

Alan F. Chalmers

5.1 Common Distinctions Between Solids, Liquids and Air

Seventeenth-century scholars could take for granted distinctions between solids, liquids and air that were by no means novel.[1] In this paper I explore the way in which these distinctions were gradually sharpened up, by way of experimentation and the theorizing accompanying it, to the point where important beginnings of a theoretical grasp of the essential distinctions between the three states of matter was achieved.

Distinctions between solids, liquids and air apparent in everyday phenomena presumably lay behind the Aristotelian theory of the four elements, air, earth, water and fire. Linked as they were to the notions of natural places and natural motions, the Aristotelian distinctions were to prove increasingly inadequate to cope with a range of phenomena revealed by experiment in the seventeenth century. As we shall see, an adequate theoretical response to such developments involved moves towards a technical notion of the concept of pressure. However, it is worth noting that as late as 1672 it was necessary for Robert Boyle to counter is some detail Henry More's resistance to his own introduction of pressure, which resistance was based on the Aristotelian presumption that water does not weigh in water.[2]

As a background to the ensuing discussion I note some knowledge of solids, liquids and air that was implicit in everyday knowledge or in technologies familiar

[1] From a modern point of view it is more natural to speak of gases rather than air, but, of course, the identification of distinct kinds of gases occurred only in the eighteenth century.

[2] Boyle's detailed rejoinder to More is in 'An Hydrostatical Discourse' in M. Hunter and E. B. Davis (eds): *The Works of Robert Boyle*, London: Pickering and Chatto, 1999, Vol. 7, pp. 141–184.

A.F. Chalmers (✉)
University of Sydney, Unit for History and Philosophy of Science, Carslaw Building F07, NSW 2006, Australia
e-mail: alan.chalmers@sydney.edu.au

F. Stadler (ed.), *Integrated History and Philosophy of Science*, Vienna Circle Institute Yearbook 20, DOI 10.1007/978-3-319-53258-5_5

by the dawn of the seventeenth century, although they were not singled out and listed in this way at the time.

Solids have a definitive shape and size, up to a point. The qualification 'up to a point' is necessary because the shape and size of solids can be modified by distorting forces. A solid normally returns to its original size and shape on the removal of a distorting force. That is, solids are more or less elastic, a phenomenon familiar from the behaviour of bows, lute strings and the expansion and contraction of the bladders of animals, for instance.

A sample of a liquid also has a natural size, but, unlike a sample of a solid, it does not have a natural shape. It adopts the shape of any container into which it is poured. Liquids flow in a way that solids do not, and solids can float in liquids but not in other solids. Unlike solids, liquids are virtually incompressible; they are inelastic. Other distinctions can be brought out by attending to differences in the behaviour of water and sand. Both can be transferred from a high to a low level via a pipe, but the water can climb a subsidiary hill in a way that sand cannot and when ejected from the bottom of the pipe it will not form a heap in the way that sand does.

A sample of air has neither a definite shape nor a definite size. It expands to fill any container into which it is put. Like solids, but unlike liquids, air is elastic. When a sample of air is compressed it resists the force of compression to a degree, as evident from resistance felt when a plunger is thrust into a closed syringe or gun barrel. The difference between the behaviour of a syringe filled with water and one filled with air exemplifies the compressibility of the latter as opposed to the former.

Phenomena suggesting that fluids press in directions other than the downwards direction of their weight were familiar by the seventeenth century. The fact that wine will be forced horizontally our of a leaking barrel and that water pushes horizontally against a lock gate, in spite of the fact that the weight of the liquids involved acts downwards, provide examples, as does the uniform expansion of a bladder when inflated which contrasts with the behaviour of a wire when stretched. However, as we shall see, a precise and quantitative grasp of the isotropy of the transmission of forces in liquids and air was only achieved as a result of seventeenth century advances. What proved to be a necessary condition for the achievement was a precise grasp of the concept of pressure.[3]

[3] The concept of pressure, while obvious to us, was not always so. The history of hydrostatics has hitherto not taken adequate account of the way in which the modern concept gradually emerged from a common sense version in the seventeenth century. There are too revealing studies that I seek to emulate of the emergence of concepts that have since become obvious. One concerns the emergence of the modern concept of motion as the result of the struggles of the likes of Descartes and Galileo to cope with the limitations of earlier concepts in Peter Damerow, Gideon Freudenthal, Peter McLaughlin and Jürgen Renn (Eds.), *Exploring the Limits of Preclassical Mechanics: A Study of Conceptual Development in Early Modern Science*, New York: Springer, 2004. The other exemplar involves the emergence of the concept of torque in the context of the balance, described in Jürgen Renn and Peter Damerow, *The Equilibrium Controversy*, Max Planck Research Library for the History and Development of Knowledge. Sources 2, Edition Open Access, 2012, http://www.edition-open-access.de.

With respect to the aim of making precise technical sense of, that is, of adequately theorizing, distinctions of the kind discussed above, there was one significant achievement that could be taken for granted in the seventeenth century. I refer to the science of weight. The latter was a mathematised theory consisting of a body of theorems derived from seemingly unproblematic axioms in the style of Euclid and Archimedes. It had been developed rigorously and in detail and applied to a wide range of mechanical systems involving balances, levers, pulleys and so on. Simon Stevin's *The Art of Weighing*, published in 1586, which included an original treatment of equilibrium on inclined planes, bears witness to the degree of sophistication reached in the science of weight of which seventeenth-century scholars were able to take advantage.[4] Another technical achievement that could be drawn on in the seventeenth century was Archimedes' theorization of the phenomenon of floating.

I conclude this introductory section by making an observation that sets the scene for my discussion. Solids and liquids alike possess weight, and a few decades into the seventeenth century it was generally appreciated that air does too. Consequently, if sciences concerning the behaviour of liquids and air were to be developed, it was necessary to grasp theoretically the distinctive features of them other than their weight. Those are the developments that I now trace.

5.2 Stevin's Hydrostatics

A key starting point is Stevin's attempt to move beyond the science of weight with his *Elements of Hydrostatics* and its short sequel, *The Practice of Hydrostatics*. I have analysed and evaluated his theory in detail elsewhere. Here I summarise and exploit the conclusions to which I have been led.[5]

On the face of it, Stevin's hydrostatics consists of a body of theorems derived from unproblematic postulates. A not over-generous summary of the content of those theorems, sometimes referred to as Stevin's law, is as follows: Normal to any solid surface element bounded by a liquid there is a force that is equal to the weight of a prism of the liquid whose cross section is equal to the area of the surface element in question and whose height is equal to the depth of the surface element beneath the liquid surface. This claim is correct from a modern point of view if we limit ourselves to the forces in liquids stemming from their own weight.

---

[4] English translations of Stevin's *Art of Weighing* and its sequel, *The Practice of Weighing*, together with the original Dutch versions, are in E. J. Dijksterhuis (ed.): *The Principal Works of Simon Stevin, Volume 1*, Amsterdam: Swets and Zeitlinger 1955, pp. 97–347.

[5] English translations of Stevin's works on hydrostatics, together with the Dutch originals are in Dijksterhuis, *Principal Works of Stevin, op. cit.*, pp. 393–501. A detailed analysis of Stevin's hydrostatics is given in Alan Chalmers, "Qualitative Novelty in Seventeenth-century Science: Hydrostatics from Stevin to Pascal" in *Studies in History and Philosophy of Science*, 51, 2015, pp. 1–10.

A close analysis of Stevin's derivation of his theorems reveals that they do not in fact follow from his postulates. Those postulates relate only to forces acting in a vertical direction, whether they are weights bearing down or the vertically acting resistances offered by liquids to them. They cannot yield the isotropic forces implicit in Stevin's law as deductive consequences. When, for instance, Stevin derives the horizontal force exerted by a liquid on a vertical plane he relies on two assumptions that he makes without signalling the fact and which are not consequences of his postulates. He assumes that the liquid does exert a force normal to a vertical plane and he assumes that the strength of that force on an element of a vertical plane at some particular depth in the liquid is equal to the force normal to an element of a horizontal plane of the same area and at the same depth.[6] In effect, Stevin, here and elsewhere in his hydrostatics, simply assumes that the forces with which a liquid presses against a solid surface are independent of the orientation of that surface.

There is no doubt that Stevin, who, amongst other things, was a hydraulic engineer *in the Netherlands*, was aware that water presses horizontally against a dyke or a lock gate and that wine issues horizontally from a leak in the side of a wine cask.[7] Such recognitions notwithstanding, the horizontal pressing of a liquid was puzzling from the point of view of the science of weight, and continued to be so for decades after 1586. How can a liquid press sideways given that its weight acts vertically downwards? The horizontal component of weight is zero! To the extent that there is a puzzle here, Stevin does nothing to help the situation by feeding in the puzzling data as an assumption, and, what is more, one that is unheralded. Stevin clearly had an intuitive grasp of the fact that liquids press equally in all directions, and this is most evident in his presentation in the *Practice of Hydrostatics*. However, he was unable to explicate this in the *Elements of Hydrostatics* because in that work he was intent on deriving his theorems from postulates that could be granted at the outset and key aspects of the behaviour of liquids went beyond what could be so granted.

Stevin's practical experience with liquids somehow guided him to hydrostatic theorems that correctly describe the phenomena. But he fell short of formulating a theory that captured mathematically the distinction between liquids and solids that would take him beyond the science of weight to a science of hydrostatics.

5.3 Limitations of Attempts by Galileo and Descartes to Characterise Liquids

In a work on hydrostatics written in 1666, occasioned by the publication of Pascal's Treatises on hydrostatics and pneumatics that we will discuss below, Robert Boyle, while expressing admiration for Stevin's hydrostatics, complained that he had

[6] Stevin's, admittedly mathematically ingenious, deduction is in Dijksterhuis, *Principal Works of Stevin, op. cit.*, pp. 421–423. The details of my critique are in Chalmers, "Qualitative Novelty" *op. cit.*, pp. 3–9.

[7] Stevin explicitly discussed the force exerted on a lock gate. See Dijksterhuis, *Principal Works of Stevin, op. cit.*, p. 497.

shown *that* his theorems are true without explaining *why* they are true.[8] For several decades following the publication of Stevin's hydrostatics, which was available in Latin from 1608, there was no significant progress towards a theory that would explain why Stevin's theorems are true. I illustrate my point with some brief comments on the efforts of Galileo and Descartes in that respect.

Galileo's key contributions to hydrostatics appeared in his work *Bodies that Stay atop Water or Move in it*, published in 1612.[9] In it Galileo analysed floating by generalising a principal that he had employed in his mechanics and which he attributed to Aristotle. In the context of a balance it asserts that if it is tilted from its equilibrium position, the balancing weights acquire velocities that are inversely proportional to those weights. Applying this to floating bodies, Galileo argued that a body will float in water in a vessel when it is at a depth such that the velocity of a small vertical displacement of the body times its weight will be equal to the velocity of the liquid that is consequently raised or lowered times its weight. He applied a similar treatment to deal with what became known as the hydrostatic paradox, where a small amount of water in a narrow tube can support a greater amount of water in a wider tube when the water in their lower parts is in communication in a closed vessel. In both instances, Galileo's treatment allows the correct equilibrium conditions to be proven. However, that treatment can be found wanting in a number of respects. While Galileo's principal yields a proof of balance conditions once they are known, no account is given of how balance arises as a result of the forces at work. What is more, the considerations are restricted to the vertical raising and lowering of weights. Galileo's hydrostatics did not move beyond considerations of weight and was ill-equipped to cope with forces acting in directions other than the vertical.

There is no evidence that Galileo was aware of Stevin's hydrostatics when he composed his own version that was published in 1612. This does not apply to Descartes. He wrote a manuscript in 1618 in response to queries put to him by Isaac Beeckman about what to make of some of Stevin's theorems and proofs.[10] Descartes addressed what he and Beeckman saw to be a deficiency in Stevin's explanation of hydrostatic equilibrium by invoking corpuscular mechanisms. The extent to which liquids bear down on the bases of containers of various configurations was explained by appeal to chains of corpuscles each one pushing on and pushed by its neighbours, a strategy that Beeckman himself had adopted.[11] This was to remain a feature of Descartes' hydrostatics, whether in the context of liquids having weight, as in wine

[8] Robert Boyle: "Hydrostatical Paradoxes Made out by New Experiments" in *The Works of Robert Boyle, op. cit.*, Vol. 5, pp. 207 and 236.

[9] A recent English translation of that work is in S. Drake: *Cause, Experiment and Science*, Chicago: University of Chicago Press 1981.

[10] Descartes' manuscript is in C. P. Tannery (ed.): *Oeuvres de Descartes, Vol. 10*, 2nd ed., Paris: Vrin 1974–86, pp. 67–74. It is analysed and its significance assessed in S. Gaukroger and J. Schuster: "The Hydrostatic Paradox and the Origins of Cartesian Dynamics" in: *Studies in History and Philosophy of Science*, 33, 2002, pp. 535–572.

[11] Beeckman's efforts to elucidate Stevin's hydrostatics in corpuscular terms are discussed in Klaas Van Berkel, *Isaac Beeckman on Matter and Motion*, Baltimore: The John Hopkins University Press 2013, pp. 134–135.

bearing down on the base of a vat, or in the context of his strict mechanical philosophy in which weight was to be explained and in which the key mechanism was the tendency of corpuscles to move away from the axes of the vortices in whose motion they participated. One problem derived from the speculative and ad hoc nature of the corpuscular mechanisms invoked by Descartes. Another stemmed from the fact that the forces arising from the action of weight or a tendency to recede from an axis or centre are directed forces ill-adapted to cope with phenomena that involve liquids pressing sideways and upwards as well as downwards.[12]

The common recognition that liquids press sideways posed a problem for those seeking to theorize or mathematize hydrostatics by extending the science of weight. But common phenomena, such as the force on a lock gate and the issuing of wine from a hole in the side of a vat, attested to the fact that liquids do indeed thrust sideways whatever the challenge this posed for theoreticians. This is encapsulated in a story told by Evangelista Torricelli, in a letter to Michelangelo Ricci, involving an interchange of a philosopher with his servant. The servant, who was engaged in fixing a faucet in the side of a wine barrel, was mocked by the philosopher for wasting his time, given the fact that the weight of the wine in the barrel presses vertically downwards. The servant persisted nevertheless and, of course, the wine did indeed flow horizontally from his faucet.[13]

Whether adequately theorized or not, the efflux of liquids from containers was explored experimentally in the early 1640s, with Galileo's successors in Italy, including Torricelli, centrally involved.[14] The speed of liquid ejected horizontally, and even vertically, from low down in a container was measured by weighing the amount of liquid collected in a given time. The results gave support to the law stating the proportionality of the square of the velocity of efflux and the depth of the orifice beneath the liquid surface. In so far as this was theorized, it was done by likening the passage of the liquid from the container to the fall of an object either freely or down an inclined plane, where the said proportionality was commonly presumed to apply. Once again, the theorization was firmly anchored in considerations involving weight.

[12] The inadequacies of Descartes' efforts in this context have been identified and discussed in Alan Shapiro, "Light, Pressure and Rectilinear Propagation: Descartes' Celestial Optics and Newton's Hydrostatics" in *Studies in History and Philosophy of Science* 5, 1974, pp. 239–296, especially pp. 240–241 and 509–514 and in John Schuster, *Descartes Agonistes: Physico-Mathematics, Method and Corpuscular Mechanism*, Dordrecht: Springer 2013, pp. 117–9 and 509–562.

[13] Torricelli's letter is in G. Loria and G. Vassura (eds.), *Opere di Evangelista Torricelli,* Faenza: Montanari 1919, Vol. 3, pp. 198–201. An English version is in I. Spiers and A. Spiers: *The Physical Treatises of Pascal*, New York: Columbia University Press 1937, pp. 167–170. All my references to Torricelli's letters in what follows are to the English versions.

[14] This work is described in C. S. Maffioli, *Out of Galileo: The Science of Waters, 1628–1718*, Rotterdam: Erasmus Publishing 1994, pp. 71–89.

5.4 Understanding Air as Distinct from Liquids and Solids in the Wake of Torricelli

Attempts to get a theoretical handle on the distinctive features of liquids and air moved forward significantly in the wake of Torricelli's experiment in 1644 involving what later became recognized as the mercury barometer. Torricelli's experiment, and, just as important, his theoretical reflections on it, were discussed in an interchange of letters between himself, in Florence, and Ricci, in Rome. Ricci sent extracts of these letters to Marin Mersenne and their contents became widely known and discussed, especially in France. These, and ensuing, discussions, especially amongst the Mersenne group in Paris and involving Blaise Pascal, were to set the scene for Pascal's moves towards significant advances in the theorization of liquids.

Torricelli, in a letter of June 11, 1644, put his experiment in the context of the assumption that we 'live submerged at the bottom of an ocean of the element air, which by unquestioned experiments is known to have weight'.[15] The mercury in the tube is held up because 'on the surface of the liquid which is in the basin, there gravitates a mass of air fifty miles high'.[16] If mercury were to be replaced by water, then the water will rise 'as much further than the quicksilver rises as quicksilver is heavier than water'.[17] He elaborated on his explanation in a letter of June 28 in response to objections from Ricci in a letter of June 18.

One of Ricci's objections involved placing a metal cap over the surface of the dish of mercury in which the barometer tube was held. He surmised that the mercury level would not fall following such an intervention, thus posing a challenge to Torricelli's theory. In part of his response, Torricelli assumed that the metal cap be placed in such a way as to leave a layer of air between it and the mercury. He argued that this layer of air, which on his hypothesis was compressed by the weight of air above it, would remain compressed after the placement of the metal cap. It would therefore continue to press on the mercury as before and the mercury level in the barometer should remain unchanged. Torricelli developed his argument by drawing an analogy between the air he presumed to weigh down on the mercury and a cylinder of wool pressing on the base of its container by virtue of its weight. He invited Ricci to imagine that a sheet of iron be inserted part way up the wool so as to isolate the bottom portion from the top portion. Torricelli insisted that, since the bottom portion of wool would remain compressed, it would press down on the base as before. 'Try it yourself', wrote Torricelli, 'for I shall not continue to bore you'.[18]

A second objection put forward by Ricci directly introduced the issue of the direction of hydrostatic forces. Ricci observed that Torricelli's explanation of the barometer assumed that the atmosphere presses down on the mercury in the dish. However, the resistance a syringe offers to the movement of a plunger is independent

[15] Spiers and Spiers, *op.cit.*, p. 164.

[16] *Ibid.*, p. 165.

[17] *Ibid.*

[18] *Ibid.*, p. 169.

of the direction in which the syringe points, so, wrote Ricci, 'it is still not evident that one can easily imagine how the weight of the air [which acts downwards] has anything to do with the effect'.[19] Torricelli invoked two empirical effects by way of a response. One involved the observation that wine will spurt out in all directions from holes deep in the side of a barrel, showing that 'although by nature liquids gravitate downwards, they press and spout in every direction, even upwards, as long as they find places to reach, – that is, places which resist with less force than their own'.[20] The second involved the observation that if a pitcher filled with air is immersed mouth-downwards in water and then a hole is made in its base to allow the air to escape, then the water moves up into the pitcher, in spite of its natural tendency to fall. Note that, here, Torricelli answers Ricci's objection, which involved the isotropy of forces in air, by examples involving isotropy in liquids! A notable feature of the developments in the decade separating Torricelli's experiment and Pascal's composition of his *Treatise on the Equilibrium of Liquids* is the extent to which the discussion moves freely between liquids, solids and gases exploiting various analogies between them.

Torricelli's theoretical reflections and arguments were highly suggestive and fed productively into subsequent developments. But they are not all of a piece. The 'ocean of air' analogy with which Torricelli begins has an appeal if air is likened to a liquid, as is directly suggested by use of the term 'ocean'. It was well known that columns of liquid in contact via their bases will be in equilibrium if their heights are proportional to their densities, the diameters of the columns being irrelevant. The barometer can be understood in these terms if the (very high) column of air resting on the mercury in the dish is likened to a column of a very rare liquid. Insofar as hydrostatic equilibrium can be comprehended in terms of balancing weights, so can Torricelli's experiment via the 'ocean of air' metaphor.

Other strands in Torricelli's reflections take him beyond an emphasis on weight. When he likened the atmosphere to a compressed pile of wool, air was likened to an elastic solid rather than a liquid. When a layer of air immediately above the mercury in the dish is cut off from the air above it, it continues to press on the mercury just as a compressed elastic solid would. The weight of air drops out of the picture and the forces engendered by its compression take its place. It is the compressed air pressing against the mercury in the dish that supports the column of mercury in the inverted tube. It may well have been the case that that air became compressed by the action of the weight of atmospheric air pressing on it from above, but the same effect would arise should the air be compressed in some other way.

The preceding observations were implicit in Torricelli's talk of the compression of low-lying air but not made explicit to the extent that I have done. We have seen that Torricelli compared compressed air to compressed solids such as wool to help make his position intelligible. However, too close an analogy between compressed

---

[19] *Ibid.*, p. 167.

[20] *Ibid.*, p. 169. Note that already Torricelli talks of the pressing and spouting in all directions as resulting from a propensity that liquids possess of 'their own', and which is distinct from their natural tendency to gravitate downwards.

air and compressed solids suffer from the limitation that the effects of compression in air is isotropic in a way that they are not in the case of solids. When Ricci, in effect, pointed to the isotropy of the forces engendered by the expansion or compression of air, pointing out that the effects of shifting the plunger in a syringe do not depend on its orientation, Torricelli, in his reply, dropped the analogy with compressed wool and invoked examples involving isotropy in liquids, as we have seen.

Freeing the question of the forces exerted by air as a result of its compression from considerations of its weight was facilitated by experiments that built on Torricelli's and which exploited the space above the mercury in a barometer. One of them, versions of which were conducted independently by Etienne Noel and Gilles Personne De Roberval in 1647, involved introducing equal volumes of water and air into that space, with dramatically different results. The introduction of the air resulted in a much greater reduction in the height of mercury in the tube than that caused by the introduction of an equal volume of water, in spite of the fact that the water weighed around a thousand times more than the air did. Roberval's discussion of the experiment led him, not only to free considerations of the force arising from the compression of air from weight considerations, but also to distinguish the character of those forces from those arising from the compression of solids. Indeed, Roberval in effect manages to capture with considerable precision key ways in which air, as such, differs from liquids and solids, the key one being the its propensity to fill any space available to it and to exert a force on a surface that stands in the way of it doing so, whatever its orientation.

What is conveyed by Roberval's text is that it is a natural property of air, that is, a property it possesses by virtue of being air, to expand spontaneously into any space available to it. When a volume of condensed air from the atmosphere was introduced into the space above the mercury 'it spontaneously and of itself became rarefied in the tube' and again, 'as a matter of fact, if besides mercury or water, there be admitted into any part of the tube some of our compressed and condensed air, as we have stated above, this air obtains its freedom and all its parts recoil and become rarified and drive out the mercury or water, which for that reason will be depressed below the aforesaid height, either more or less, according to the air itself possesses greater or lesser power of rarefaction'.[21] Air has a 'power of rarefaction' not possessed by liquids or solids.

The detachment of the force exerted by condensed air from weight considerations was taken a step further by Roberval's insertion of a carp's bladder, freed of most of its air and tied at the neck, into the space at the top of the barometer. The force of expansion exhibited by air was illustrated in a visually compelling way when the bladder was seen to expand markedly.[22]

---

[21] The quotations are from the English translations of Roberval's letters in C. Webster, "The Discovery of Boyle's Law and the Concept of the Elasticity of Air in the Seventeenth Century" in: *Archive for the History of Exact Sciences* 2, 1965, pp. 497 and 499. This reference gave me an invaluable starting point for the material discussed in this section.

[22] See *ibid.*, p. 496.

As well as supporting the view that air exerts an expansive force that is greater or less depending in its degree of compression, this experiment gives a visual display of the isotropy of the expansive force, the near spherical shape of the bladder arising from the fact that it is 'pressed by force on all sides'.[23] This remark of Roberval's reinforces the point he had already made in the context of the introduction of a small volume of air into the Torricellian space, where he had stressed that the introduced air presses 'in all directions on the adjacent bodies', driving the mercury downwards because it is only it, rather than the containing glass, that can give way.[24]

The point that air possesses a power of expansion which distinguishes it from liquids and which is implicated in the new experiments we have described above, as well as in the new understanding of the phenomenon of atmospheric pressure, was expressed very forcefully by J. Pecquet. As C. Webster has noted, Pecquet was an anatomist who became interested in the new experiments on air and, in 1651, included his own account and interpretation of them in a book on anatomy.[25] Pecquet introduced the term 'elater' to single out the propensity of air to exert a force arising from its degree of compression, and insisted that it was this feature of air that distinguished it from liquids. 'So', wrote Pecquet in the context of the experiment that compared the effects of introducing water and air into the barometer tube, 'it is evident that the adjacent mercury was not forced down as much by weight as by the strongest elater; and it is thus that the Terraquaceous Globe is compressed by air'.[26] Here Pecquet explicitly distinguishes the weight and elator of air.

5.5 Pascal's Moves Towards a Technical Concept of Pressure

Pascal was fully aware and able to take advantage of the developments we have discussed and participated in them himself, performing experiments of his own. In 1647 he published an account of his early experiments that included a version of Torricelli's experiment using water in place of mercury.[27] He orchestrated a repetition of Torricelli's experiment at the top of the Puy de Dôme, conducted in 1648. My focus in this Section is on the drawing together of all of these considerations in the two Treatises that Pascal wrote in 1654, *The Equilibrium of Liquids* and *The Weight of the Mass of the Air*. In the first of them in particular, we find an advance in the characterization of the distinguishing features of liquids which was to be a significant move towards the modern concept of pressure.[28]

---

[23] *Ibid.*, p. 497.

[24] *Ibid.*, p. 498.

[25] See Webster, *op. cit.*, pp. 451–454.

[26] *Ibid.*, p. 499.

[27] B. Pascal, "Experiences nouvelles touchant le vuide" in L. Brunschvigg and P. Boutroux (eds.): *Oeuvres de Blaise Pascal, Volume 2*, Paris 1908, pp. 74–76.

[28] Pascal's treatises were published posthumously in 1663. My references are to the English translation in Spiers and Spiers, *op. cit.* I have discussed Pascal's theory in more detail in Chalmers,

In the subtitle of his Treatises Pascal promised to provide 'the explanation of the causes of various effects of nature which had not been known hitherto'.[29] As far as liquids are concerned, he provided explanations of a range of hydrostatics effects, some of them novel and some of them long known, by describing how forces applied to a liquid, whether they originate from weights or humans pushing on a piston or from the weight of the liquid itself, are transmitted through the body of the liquid. Pascal specified that the force per unit area at the point of application of a force to a liquid in a container appears at any other surface of the liquid, whatever its orientation, as the same force per unit area.[30] This behavior was attributed by Pascal to features characteristic of liquids as such, namely, their 'continuity and fluidity'.[31] This is borne out by the fact that hydrostatic effects are destroyed if the liquid responsible for them is frozen.[32] Pascal articulated his account of hydrostatics by reference to a set up that was to become known as the hydraulic press. He stressed the point that the multiplication of force achieved by such a press comes about because the water 'is equally pressed upon under the two pistons', the force being greater on the larger one in proportion to the degree to which its cross sectional area exceeds that of the smaller one.[33] This constituted a significant move towards the modern concept of pressure but fell short of it in ways that I identify in Sect. 5.6.

We have seen that the path of the previous decade that led to Pascal's hydrostatics involved interplay between the exploration of liquids and air and analogies between them both and the behavior of solids, with novel conceptualizations as well as novel experimental findings emerging. Pascal did not take maximum advantage of those developments when it came to his second treatise, *The Weight and Mass of the Air*. The title itself hints at what was indeed the case, namely, that Pascal was intent on explaining pneumatic phenomena by appealing to the weight of the air and, as a consequence, failed to match the progress he made with the distinction between solids and liquids with a theorization of the way in which air differs from both of them. As I have stressed above, air does not differ from solids and liquids by virtue of possessing weight.

In the first chapter of *The Weight and Mass of the Air* Pascal spells out consequences of the recognition that air has weight and sets out his pneumatics in subsequent chapters by appeal to those consequences. In particular, he is intent on showing that phenomena thought to require appeal to the force of a vacuum for their explanation can all be explained better by appeal to the weight of the air. He notes that atmospheric air presses on the earth by virtue of its weight just as the oceans do and he observes that the degree of this pressing is proportional to the height of air,

---

"Qualitative Novelty", *op cit.*, pp. 6–8, although my critique of Pascal's Treatise on air is an addition to what is covered in that place.

[29] Spiers and Spiers, *op. cit.*, p. xxix.

[30] For the clearest formulation by Pascal of this latter, quantitative, point see Spiers and Spiers, *op. cit.*, pp. 7–8.

[31] Spiers and Spiers, *op. cit.*, pp. 7–8.

[32] Spiers and Spiers, *op. cit.*, pp. 4 and 10.

[33] Spiers and Spiers, *op. cit.*, p. 7.

just as is the case with water. He also observes that 'bodies in the air are pressed on all sides by the weight [*le poids*} of the air above them', drawing a comparison with what he had shown in *The Equilibrium of Liquids*.[34] The reference to 'pressing on all sides' alludes to the isotropic character of the pressing but the centrality of that notion is obscured by the use of the term 'weight', a feature that is characteristic of the whole of Pascal's treatise on air.

By developing his pneumatics by emphasizing the weight of air, and also exploiting isotropy of the forces exerted by liquids on solid surfaces only covertly, Pascal had sidestepped the issue of just what it is that distinguishes air from liquids and solids. He failed to exploit to full advantage experimental findings and their attempted theorization by the likes of Torricelli, Roberval and Pecquet. Air shares with liquids and solids the property weight, and it shares with liquids, but not with solids, the isotropy of the distribution of forces within it. It also shares with solids a certain kind of elasticity insofar as it responds to contractions with a resisting force, although there are differences insofar as the elasticity of air is isotropic and air has a spontaneous propensity to expand and so does not offer a resistance to expansion in the way that solids do. These differences were already illustrated, by experiments performed by Roberval for example, as we have noted. Key ones involved the effects of introducing a small volume of air into the Torricellian space, as compared to introducing an equal volume of water, and the effect of placing an airtight bladder largely free of air into that same space. Such experiments are simply not mentioned in Pascal's treatise on air. Pascal contributed significantly to the task of identifying the distinguishing features of solids and liquids but he did not succeed in capturing the distinctive feature of air with equal clarity. Pascal did not make significant progress in that respect because he failed to capitalize on the grasp of the character of air that Roberval had alluded to by his 'power of rarefaction' and Pesquet had with his concept of elator.

5.6 Beyond Pascal to Boyle's Concept of Pressure in Liquids

In recent articles I have followed Pierre Duhem and Dijksterhuis and credited Pascal with having formulated the modern concept of pressure.[35] I now consider that claim to be in need of serious qualification. According to its modern usage, 'pressure' specifies the state of a fluid throughout its volume. A fluid presses equally on either side of any plane within a body of fluid in equilibrium. Within the body of a fluid the net force due to this uniform pressing in all directions is zero. It is only on solid

[34] Spiers and Spiers, *op. cit.*, p. 29.

[35] See Alan Chalmers, "Intermediate Causes and Explanations: The Key to Understanding the Scientific Revolution" in *Studies in History and Philosophy of Science* 43, 2012, pp. 551–562, especially pp. 555–557, Chalmers, "Qualitative Novelty", *op. cit.*, pp. 5–8, P. Duhem, "*Le Principe de Pascal*" in: *Revue Gènèral des Sciences Pures et Appliques*, Vol. 16, 1905, pp. 599–610 and Dijksterhuis, *Simon Stevin, op. cit.*, p. 69.

surfaces bounding a fluid that the pressure is no longer balanced and a net force results that is equal to the product of the pressure and the area of the surface. Pascal identified the forces on solid surfaces, but as far as the transmission of forces through the body of liquids is concerned he attributed it to their 'fluidity and continuity' without further elaboration. In his Treatise Pascal talked freely of liquids pressing on surfaces using the verb '*presser*' but he did not use the noun '*la pression*'. A full account of how hydrostatic effects are caused must include the details of how forces are transmitted through liquids by way of elements of liquid pushing against and being pushed by their neighbours. Insofar as Pascal fell short of doing that, he fell short of capturing the notion of pressure that is necessary to specify the state of a liquid and that is distinct from weight.

It was left to Robert Boyle to modify and extend Pascal's account in order to supply what was needed in these respects.[36] Boyle introduced the theoretical device, now taken for granted, of considering the forces acting on either side of imaginary planes in the body of a liquid. He thereby specified the way in which forces are transmitted through liquids and so provided a full mechanical account of how hydrostatic effects are brought about. Boyle did employ a concept of pressure that was an anticipation of the modern one and made full use of the term 'pressure' to describe it. It was also left to Boyle to capture the distinctive feature of air by way of the 'spring of the air', his elaboration of Roberval's 'power of rarefaction' and Pecquet's 'elater'. I have recently analysed these contributions of Boyle in detail and will not reproduce them here.[37]

---

[36] Boyle's "Hydrostatical Paradoxes" *op. cit.*, is a commentary on and elaboration of Pascal's *The Equilibrium of Liquids.*

[37] See Alan Chalmers, "Robert Boyle's Mechanical Account of Hydrostatics and Pneumatics: Fluidity, the Spring of the Air and their Relationship to the Concept of Pressure" in *Archive for History of Exact Sciences* 29, 2015, pp. 429–454.

Chapter 6
"Beyond the Conventional Boundaries of Physics": On Relating Ernst Mach's Philosophy to His Teaching and Research in the 1870s and 1880s

Richard Staley

Abstract Ernst Mach's most well known critiques of mechanics concern mass, inertia and space and time. Conceptually motivated towards avoiding unnecessary assumptions and basing physical concepts on measured relations, they were first published in the years around 1870 (for mass and inertia) and in his well known 1883 book *Die Mechanik in ihrer Entwickelung historisch-kritisch dargestellt*, later translated as *The Science of Mechanics: A Critical and Historical Account of its Development*. Philosophical discussion of Mach's critiques has reflected these conceptual concerns, connecting them to Mach's account of science as the economical description of phenomena. Yet manuscript records of his teaching in the 1870s show that Mach was also animated by psychophysics and the relations between inner and outer worlds. His publications attest to these broader interests as well. In the 1870s, for example, Mach developed physiological studies of the sense of motion. Soon after completing his critical history of mechanics he took up the relations between physiology and psychology in his 1886 *Beiträge zur Analyse der Empfindungen*. By investigating Mach's research across subject matter that has usually been treated separately, and integrating his teaching with his research, this chapter aims to offer a study of Mach's philosophy as it is revealed in practice. Mach presents a highly unusual example of someone whose primary aim was to reform his own discipline of physics through the concerns of other disciplines, something he alluded to in 1886 when stating that he expected the next great enlightenments of the foundations of physics to come at the hands of biology.

R. Staley (✉)
Department of History and Philosophy of Science, University of Cambridge,
Free School Lane, Cambridge CB2 3RH, UK
e-mail: raws1@cam.ac.uk

F. Stadler (ed.), *Integrated History and Philosophy of Science*, Vienna Circle Institute Yearbook 20, DOI 10.1007/978-3-319-53258-5_6

6.1 Beyond the Conventional Boundaries of Physics

In the preface to his difficult and quirky 1886 book *Contributions to the Analysis of the Sensations*, Ernst Mach noted that he expected the next great enlightenments of the foundations of physics to come at the hands of biology, and specifically from the physiological and psychological analysis of sensations that was the subject matter of his book. Mach wrote that he had no claim to the title of physiologist and still less to that of philosopher, but instead described himself as a physicist who was "unconstrained by the conventional barriers of the specialist" (Mach 1897, vii–viii). Erik Banks has provided an excellent and comprehensive study of Mach's philosophy (Banks 2003). More recently Ed Jurkowitz's 2009 unpublished manuscript "Liberal Pursuits: Hermann Von Helmholtz, Ernst Mach and the Framing of Physics and the Human Mind," and researchers like Paul Pojman and Alex Hui have begun to show how different elements of Mach's research in sense physiology may be related to his general philosophical approach (Pojman 2011a, b; Hui 2013). Yet the dominant image of Mach is probably still that of the positivist anti-metaphysician, based largely on his contributions to physics and summarised by the thought that sensations provide the basis for all knowledge and the aim of science is the most economical description of them (Holton 1993; Banks 2012). In this chapter I want to suggest that whether or not he was right about the future significance of biology, Mach's 1886 comment offers a revealing portrait of his own work up to that time. I will argue that Mach's conceptual approach to the foundations of physics was more deeply shaped by physiology and psychology than we have realised, and in any case his research in all these fields was closely related. I will support this argument by studying Mach's research and teaching in the 1870s and 1880s, trying where possible to set his published papers in the light of the laboratory notebooks and lecture manuscripts that remain in the Deutsches Museum. As well as integrating both teaching and research, and reaching across the span of Mach's interests from psychology to physics, this approach will provide a good means of exploring the relations between Mach's philosophy and his research as it developed in practice. In these diverse senses I hope to contribute to an integrated history and philosophy of science.

6.2 Mach's Early Career

After completing his doctorate on electrical discharge and induction in Vienna in 1860, Mach began his career with lectures on physics for medical students. His research traced physical phenomena between motion and the senses, and the mind and the senses, absorbing but also critiquing what Gustav Fechner called psychophysics (Blackmore 1972). Like Hermann von Helmholtz, Mach was one of many who moved between physiology, physics and psychology. Examining at the same time physical phenomena, the senses that perceived them, and aesthetics, they were

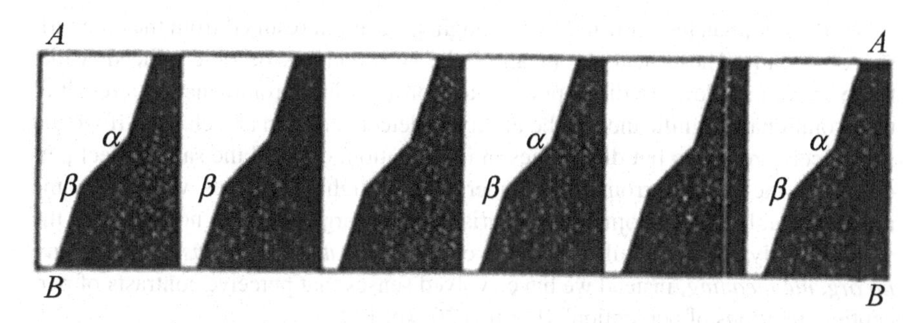

Fig. 6.1 Mach's illustration of the technique used to elicit the phenomena now known as Mach bands. When this strip is pasted on a cylinder and spun, bright and dark bands appear at α and β respectively (Mach 1886/1897, Fig. 27)

knitting together the subject matter of three laboratory sciences that were beginning to burgeon institutionally.

Several elements of Mach's research in this period informed both his rather complex philosophical stance towards sensations as the basis of knowledge, and his later critiques of Newton. Firstly, Mach was intimately concerned with how the relative motion of a source and the observer affected perception of sound or light (Mach 1860, 1873a). It is notable that in these cases observed differences in perception could be resolved by a better understanding of the physical phenomena, as long as one recognised the critical role of relative motion. You do not need to know much about the ear or eye to understand the shift in tone or wavelength that results from the Doppler effect, for example. At this time Mach also pursued a psycho-physical parallelism and so expected to find that some effects that seemed largely psychological would actually be found to have a physical correlate. One is the very different perception we have of a chord change if we concentrate on the high or low note when the first chord is sounded, a phenomena called accommodation. As Alex Hui has shown, Mach thought that shifting attention in this way would lead to changes in the ear, making the physiology of the ear critical to understanding what we hear in the different instances. Yet nearly a decade's research failed to disclose such a physiological basis. Hui argues that this led Mach to focus on the importance of distinctively historical explanations, coming to believe that significant features of our hearing had to be learned; they depended upon cultural development (Hui 2013, 89–121).

Other phenomena taught something very different. In 1865 Mach discovered what we now know as Mach bands. The set up is illustrated in Fig. 6.1. If the strip of paper illustrated is pasted on a cylinder and spun rapidly, the physical phenomena might lead you to expect that we would see a varying field of black and greys with two sharp changes. In addition, however, we see two bands, a bright line at α and a dark one at β (Mach 1886/1897, Fig. 27; Ratliff 1965). In this case Mach had disclosed an effect that was physiological rather than being either simply physical or dependent on psychological attention or cultural learning, and his explanation was both relational and evolutionary.

As Paul Pojman has shown, Mach thought perception resulted from the comparison of stimuli. Those near the mean of the surroundings become effaced, while those above or below are disproportionately brought into prominence as a result of the evolutionary significance of the ability to detect variation and change. If we did not perceive relations but differences in illumination, one and the same object perceived in the same surroundings under diverse light intensities would become unrecognizable. Thus, Pojman summarises, Mach argues we do not perceive the world directly, for that would amount to chaos: "*Sensations by themselves can have no organic meaning*, instead we have evolved senses that perceive contrasts of perception, relations of perception" (Pojman 2011b, 124).

There are two important orientations to recognize in this early work on motion and sensation, and the mind and the senses. Firstly, Mach encounters such diverse relationships between physiological, psychological and physical phenomena that he hesitates to explain one in terms of the other. Secondly, he regards relational perspectives as critical and often explores them through studies of motion. A further significant point emerges from a paper that Mach published on the development of conceptions of space in 1866. Here, building on Hering's discussion of visual space, Mach distinguishes between physiological spaces determined by sensory perception, and geometrical spaces, which are metrical, depending on measure. And he identified a third form of space, physical, which depended on the forces of matter (Mach 1866). This kind of distinction between interrelated perspectives on phenomena was to be characteristic of Mach's approach. Thus Mach's later focus on sensation, relative motion and space has highly concrete but also complex and subtle grounds in his early studies of sense perception.

6.3 Mach's Anti-metaphysics

Even when Mach began developing a distinctly anti-metaphysical approach to physics in 1867, a revealing trace of this broader framework to his thought remains, although it has usually escaped notice. Focusing on the definition of mass, Mach aimed to clarify the distinctions between a priori, empirical and hypothetical elements. This was a much more abstract and philosophical study, and it is seen as a foundational argument for the operationalism that Percy Bridgeman helped make famous much later. But we should note that Mach argued first that the only feature we can take to be a-priori is the law of cause and effect, and specifically comments that whether that depends on a powerful induction or has its ground in our psychic organization can be left undecided, because in the psychic life also, similar effects follow similar causes (Mach 1868, 355). He then went on to offer a ground-breaking definition of mass solely in terms of the mutual acceleration of different bodies, describing this as an attempt to improve on Newton's formulation by developing a completely scientific treatment of the fundamental laws of mechanics. The fact that Mach's short paper was rejected by the prestigious *Annalen der Physik* was telling, and this reinforced Mach's caution about discussing such ideas with physicist colleagues (it eventually appeared the following year in *Carl's Reportorium*).

Four years later Mach developed this line of argument further in an 1871 lecture to the Royal Bohemian Society of the Sciences on the root and history of the conservation of work, which was published a year later with a set of footnotes that took up his critique of mechanical laws (Mach 1911). Manuscripts of his lectures at the University of Prague in the same period offer a still more pointed indication of the intellectual breadth of Mach's approach, as well as showing the fascinating methodological stance he advocated for his students (Mach 1872). I will first discuss Mach's thought on mechanics, then turn to his general philosophical perspective and critical methodological suggestion, before finally considering his understanding of the ego.

6.4 Mass and Mechanics

Mach took the opportunity of the footnotes to his lecture to reprint his discussion of the definition of mass, but also to set out an argument that Einstein later christened Mach's principle. The central point Mach made was that Newton's first law of inertia was undefined without specifying the actual bodies in relation to which a given body remained at rest or in uniform motion. This usually meant the laboratory, or the fixed stars, and Mach insisted that just because we can usually substitute another room or reference star didn't mean we could abstract to an idea of motion in absolute space, independent of the particular material bodies through which we always actually form our understanding of coordinate systems. He made his point concrete by discussing rotation. Geometrically there is no difference between whether we consider the earth at rest and the stars spinning around it, or consider the stars to be fixed and describe the rotation of the earth, the relative motion is the same. But the second case is simpler astronomically, and in the ordinary conception of inertia it has a consequence that the first does not: a rotating body is subject to the inertial, centrifugal forces that lead the earth to bulge around the equator, and that explain the complex motions of Foucault's pendulum. To solve this difficulty you could either consider all motion to be absolute, as Newton and others had done, or recognize that the law of inertia is wrongly expressed. Mach took the latter option, suggesting that what is required for the description of a phenomena should be regarded as part of the causal nexus, and arguing that we have to ask what share every mass has in the determination of direction and velocity in the law of inertia. He thought we forget this given the stability of our normal experience, but if an earthquake were to shake the earth, or the heavens were to swarm in confusion, we would be forced to ask what would become of the law of inertia, how would it be applied and expressed. Similarly, he said, a bankruptcy makes it clear that all money is important to the funding of a paper note (Mach 1911, 75–80).

The first thing I want to emphasize about Mach's 1872 discussion is its cosmic scale, considering the masses of the universe in earthquake and star-storm and linking the behaviour of each body with the distant masses of the universe through the law of inertia. My second point concerns what Mach's economic analogy reveals about his general philosophical perspective. His notes told Mach's readers that since the beginning of his teaching he had stressed that science is chiefly concerned with the convenience and saving of thought; but a highly important corollary to this emphasis on economy was his view that the process of concept formation and abstraction was similar across all areas of thought (Mach 1911, 48, 55, 88). This is something that is often forgotten when historians and philosophers focus on Mach's insistence on measurability for physical concepts, but it comes across very clearly in the classroom lectures he gave in this period. They illustrate the range of Mach's thought and the centrality of psychophysics for his approach to physics.

6.5 Mach's Philosophy: Sensation, Matter, Soul

Manuscript notes for his 1872 lectures on the principal problems in physics show that Mach began by arguing that sensation is a general property of matter. If it is not regarded as an emergent property of groups of molecules, then it has to be considered part of the elements themselves. Mach then asked where the soul is located. Experiments with animals showed it couldn't be located in the brain: the soul isn't so simple, he said, exploring an analogy with the state, which he commented "looks like a person." Mach asked his students to consider Bismarck's soul before moving between the soul of the professions, individuals and back to the state (Mach 1872). Five years after the Prussian defeat of Austria and shortly after Bismarck had been made Imperial Chancellor of a newly unified Germany, Mach's questions made an abstract point about parts and wholes with direct political meaning, without at any point stating his own political stance. Nevertheless it is likely that Mach's politics was important to many of his readers, both attracting some readers, like those in the socialist circles around Einstein as a student in Zürich, and repelling others, like Max Planck (Feuer 1974, 11–22; Heilbron 2000, 44–60).

In this particular instance Mach's underlying philosophical point was that we attribute souls to others by adding them in thought, by analogy with our direct experience of our own outer and inner sides. Scientific concepts are no different. When he asked his assistant Hajek to bring a prism into the room, one could understand Hajek as an automaton, but that was much more difficult than ascribing him a soul; just as potential theory or Ampere's rules had a practical economical value. Mach's course was dedicated to overcoming the cleft between the physical and the psychic. He finished his introductory lecture with striking methodological advice, which is bold in its epistemological and ontological parity and deliberately philosophical, humdrum and poetic at once. He told his students:

> So in fact we can hope to come to a better understanding of the world if we measure ourselves with the standards of the outer world and the outer world with our own standards, considering it as physical process, but attributing sensation to matter (Mach 1872).

This kind of comparative inversion may seem a long way from what we know about Mach's attack on Newtonian absolutes, but in fact I think there's an interesting practical pathway from his methodological advice to the critique of Newton's bucket that Mach offered in 1883. Further, considering that critique alongside the physiological and psychological perspective that Mach described in his 1886 book on sensations, will show strong relations between Mach's thought on the outer and inner worlds.

An important illustration of Mach's methodology comes in his work on the physics and physiology of our sense of balance and the vertical. This was stimulated by unusual experiences while travelling on a train in 1873, which led Mach to conduct experiments on our sensory experience of rotation, but also of fall. To do so, Mach built a rotating frame, as well as a see-saw device and an inclined plane, and he experimented extensively on the different factors leading to our bodily awareness of motion in such circumstances. Before writing extensively on Galileo's and Newton's contributions to mechanics in 1883, then, Mach had himself experienced the controlled fall of trundling down an inclined plane in a trolley, or being swung up and down, and he had sat in a chair spinning on a central axis like a bucket on a string, or rotating like a child on a merry-go-round or train-traveller going around a curve. Paying close attention to distinguishing the role of the head, muscles, skin and the visual system in such motions and experimenting with pigeons and rabbits as well as himself, Mach established firstly that we perceive the resultant of centrifugal and gravitational forces as vertical. Secondly, he made the physiological discovery that the motion of the fluids in the inner ear provides a physical, mechanical system that allows our nerves to register the vertical and changes in orientation, and he showed that the nerves accommodated to constant forces but reacted to changes in them to give a sensitive perception of rotation (Mach 1875; Staley 2013). Mach was quite literally bringing the measures of the outer world into play in order to analyse our sense of motion.

There are some interesting reflections of this physiological research in Mach's later work on mechanics, but they are often somewhat indirect. Interleaved into his notes for his 1880 lecture course on mechanics is a brief note that our own body is indispensible in establishing the laws of mechanics (Mach 1880). His 1873 laboratory notebook shows that his research on rotation and fall led Mach to return to the history of mechanics with the intention of discussing "Newton, once more," (Mach 1873b) but in fact his first sustained new thoughts on Newton seem to have come in 1879, when he discussed the Newtonian epoch and drew a diagram that seems to prefigure his 1883 discussion of Newton's bucket (Fig. 6.2).

Remember that in 1872 Mach had discussed Foucault's pendulum and the bulging of the earth at the equator when critiquing Newton's recourse to absolute motion. Now he seems to have narrowed his analysis down to the particular example that Newton had offered, of a spinning bucket with its water, before shifting scale in a speculative indication of the difficulties of understanding how to approach the phenomena if the bucket's walls were to be expanded dramatically. In this diagram he notes first that it makes no sense to assume absolute space, and then that there can be no talk of the relative motion of the inner cylinders alone, because of the existence

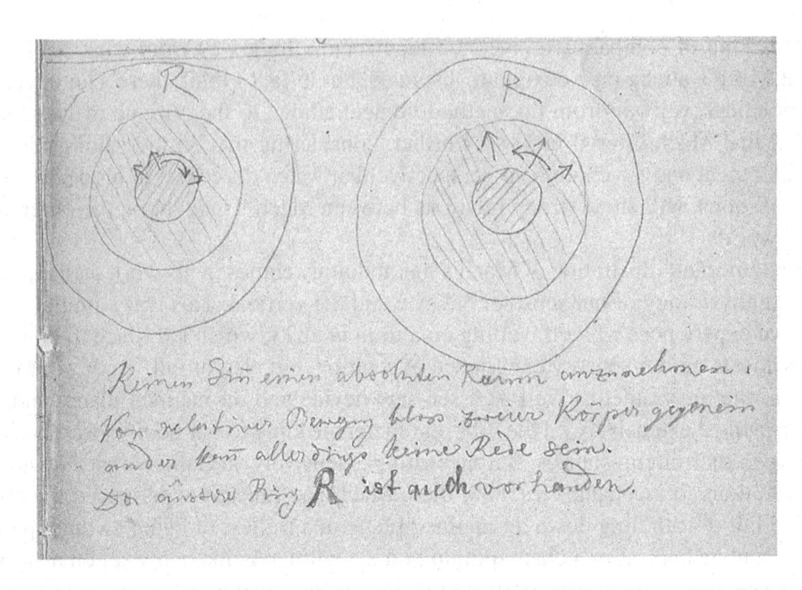

Fig. 6.2 This sketch of relative rotation anticipates Mach's published discussion of Newton's bucket experiment. Deutsches Museum, Munich, Archives (Mach 1879)

of the outer ring – presumably representing the fixed stars (Mach 1879). In 1883 he was to write "No one is competent to say how the experiment would turn out if the sides of the vessel increased in thickness and in mass till they were several leagues thick" (Mach 1883, 216–17). Here I want to note that Mach's critical thought and researches on rotation had moved from the cosmic scale of the universe to the more ordinary and intimate scale of the body and the inner ear as well as that of the bucket. It had traversed both physics and physiology.

But there was a psychological correlate to this shift in scale and move between the universe and the body, which becomes visible when Mach's approach to mass is compared with his understanding of the ego.

Mach's 1879 notes on rotation and mechanics are interrupted by passages on politics and psychology; and perhaps all of these were interrelated. One conjunction raises the possibility that Mach observed the methodological principle he urged on his students, to explore the inner world with the measures of the outer world and the outer world with those of the inner. At the very least it will suggest that Mach recognized conceptual similarities between physical and psychological conditions. Between a series of notes devoted first to action and reaction, inertia, pressure and counter-pressure, and then to siphons, Foucault's pendulum and gyroscopes, Mach wrote: "Personal impersonal. The personal lies in strong connections. The person has only apparent unity. No boundaries. Renunciation of egotism. The person is an illusion. Ethical improvement.—What one now does unconsciously one will then do consciously" (Mach 1879). Just as Mach insisted you had to consider the distant stars when understanding the centrifugal forces that result in Newton's bucket, he dissolved the boundaries between the person and the world.

Fig. 6.3 This sketch illustrates the view through one eye; Mach commented that identity came "more through the environment, than through psychic identity." Deutsches Museum, Munich, Archives (Mach 1882)

These comments have a very psychological cast, but there was a more physiologically oriented version of the point too, which is illustrated by a sketch Mach made in a later notebook, showing the view of his own body and his study as he sits with one eye closed (Fig. 6.3). His caption reads "The continuity of familiar stimuli. Identity more through the environment than through psychic identity" (Mach 1882). Mach believed neither the masses of the universe nor human bodies nor the self could be correctly understood without recognizing their relations to the environment, which in each case played a greater role in their formation than was commonly appreciated.

My intention here has been to identify significant interrelations between different aspects of Mach's thought, without investigating either where they came from or the extent to which they were taken up by others (or neglected). Yet while some elements of the inclusive understanding I have offered are reflected in the historian John Blackmore's early biography, and are pursued in Erik Banks's comprehensive study of Mach's philosophy, this kind of developmental picture of the relations between Mach's philosophy and his research has been hard to perceive. One reason is probably that philosophical attention has focused on economy and not on Mach's views about the common role of abstraction across all thought, and has neglected his methodological readiness to move between the inner and the outer. I think another reason comes down to the way Mach often presented his work. Ironically, this reflects his view about transgressing the subject boundaries of the specialist. Mach crossed such boundaries more rigorously than most scientists in this period and often emphasized the need to go beyond them, but he also thought in terms of them. Both considering as he did the different subject matters of physics, physiology and psychology, and addressing the diverse audiences of physicists, physiologists, biologists, and others, Mach was determined to reform each subject through the concerns of the other two. He always did so with a comprehensive and consistent perspective on their interrelations, but often treated these abstractly enough for his readers to miss just how intimately engaged they were in his own thought. His two major early books provide revealing illustrations of what I mean.

Mach's 1883 book on the development of mechanics begins with a discussion of the principle of the lever. Just when you think he might have finished a rather simple example of the balance, he observes that we commonly forget it relies on several assumptions. When a beam supported at its midpoint has two equal weights suspended at equal distances, they will be in equilibrium. But to be complete, Mach writes, we might note negatively that different colors on the arms, the position of the observer, nearby processes, and etc., can all be neglected (Mach 1883, 9). It is revealing that Mach prompts us to remember that we have chosen to neglect color and the observer. His book emphasized what we neglect when we focus on mechanics, and he often stressed the need to know when the observer is irrelevant or not, but its principal subject was the physics of phenomena and not those cases where physical, physiological and psychological aspects met. Thus Mach's physiological studies of the perception of motion went largely unremarked in 1883, though aspects of his experiments on fall found a place in a discussion of action and reaction that is the most proximate source for Einstein's happiest thought, and the attentive reader would come across a footnote stating that Mach would not discuss the physiological nature of time and space sensations, even as he embarked on his critique of absolute time and space (Staley 2013).

When Mach addressed the physiology and psychology of sensation in his 1886 *Contributions to the Analysis of the Sensations* he offered a worked up version of the drawing of the world seen through one eye that he had sketched as early as 1882. Now Mach used it in the context of a discussion of the relations between our bodies and other peoples' bodies, and of the domains of physiology, psychology, and physics, noting that the first two came into play when the phenomena at issue passed through the skin, so to speak. But this image also formed an introduction to Mach's discussion of the ego and the world, where Mach noted how useful our ordinary conception of the ego is, but also that the researcher should go further, to recognize that the ego is not definite, unalterable or sharply bounded. Mach cited an accord with Weismann's cell theory in arguing that continuity alone is important, and famously he concluded that the ego is unsavable (Mach 1897, 20).

For Mach, the boundaries between disciplines and between self and world are not fixed, and inertial masses, human bodies and selves are all more intimately related to the rest of the universe than is usually recognized. His insistence on this point in considering the ego was as radical as it was in considering the law of inertia; and both of these arguments stemmed from a common approach to the sciences. They should be considered alongside each other too. Doing so will change our understanding of Mach, but it may also give us new perspectives on the disciplines to which he contributed. The extent to which both contemporaries and later historians and philosophers have highlighted Mach's anti-metaphysical arguments sometimes seems to have obscured the significance of the philosophical means that Mach offered for approaching different disciplines within the same framework, as well as the methodological means he provided to pursue interrelated concerns at the same time. We need to understand why these elements were difficult to perceive, in charting the extent to which they were taken up, combatted or neglected. For example, Mitchell Ash's study of the origins of Gestalt psychology indicates particularly complex interrelations between the subject matter, methods and philosophical impetus that Mach's work provided for figures such as G.E. Müller and Wolfgang Köhler (Ash 1995). Similarly, perhaps Mach's thoughts on the ego could have served a fruitful function in the development of the philosophy of mind had they been pursued more fully, just as Heidelberger and Banks have indicated that his functional approach and neutral monism are pertinent to ongoing concerns in the study of consciousness (Heidelberg 2010; Banks 2010).

References

Ash, Mitchell G. 1995. *Gestalt Psychology in German Culture, 1890–1967: Holism and the Quest for Objectivity*. Cambridge: Cambridge University Press.

Banks, Erik C. 2003. *Ernst Mach's World Elements: A Study in Natural Philosophy*. Dordrecht: Kluwer.

———. 2010. Neutral Monism Reconsidered. *Philosophical Psychology* 23: 173–187.

———. 2012. "Sympathy for the Devil: Reconsidering Ernst Mach's Empiricism." Review of John Blackmore, et al., eds. Ernst Mach's Prague, John Blackmore, et al., eds. Ernst Mach's Philosophy: Pro and Con. *Metascience* 21: 321–330.

Blackmore, John T. 1972. *Ernst Mach: His Work, Life, and Influence*. Berkeley: University of California Press.

Feuer, Lewis S. 1974. *Einstein and the Generations of Science*. New York: Basic Books.

Heidelberger, Michael. 2010. Functional Relations and Causality in Fechner and Mach. *Philosophical Psychology* 23: 163–172.

Heilbron, J.L. 2000. *The Dilemmas of an Upright Man: Max Planck and the Fortunes of German Science. With new afterword*. Cambridge, MA: Harvard University Press.

Holton, Gerald. 1993. *Science and Anti-science*. Cambridge, MA: Harvard University Press.

Hui, Alexandra. 2013. *The Psychophysical Ear: Musical Experiments, Experimental Sounds, 1840–1910*, Transformations: Studies in the History of Science and Technology. Cambridge, MA: MIT Press.

Mach, Ernst. 1860. Ueber die Aenderung des Tones und der Farbe durch Bewegung. *Sitzungsberichte der Mathematisch-naturwissenschaftlichen Classe der Kaiserlichen Akademie der Wissenschaften Wien* 41: 543–560.

———. 1866. Bemerkungen über die Entwicklung der Raumvorstellungen. *Zeitschrift für Philosophie und philosophische Kritik* N.F. 49: 227–232.

———. 1868. Ueber die Definition der Masse. *Repertorium für physikalische Technik, für mathematische und astronomische Instrumentenkunde (Carl's Repertorium der Physik)* 4: 355–359.

———. 1872. *Manuskript Vorlesung, Ueber einige Hauptfragen der Physik, 1872 Sommer*. Deutsches Museum, NL174/449 [Mach 1872].

———. 1873a. *Beiträge Zur Doppler'schen Theorie Der Ton- Und Farbenänderung Durch Bewegung. Gesammelte Abhandlungen*. Prag: J.G. Calve.

———. 1873b. Notizbuch, ab 12.4.1873. Deutsches Museum, NL174/507 [Mach 1873(b)].

———. 1875. *Grundlinien der Lehre von den Bewegungsempfindungen*. Leipzig: Engelmann.

———. 1879. Notizbuch, ab 25.10.1879. Deutsches Museum, NL 174/519 [Mach 1879].

———. 1880. Manuskript Vorlesung. Experimentalphysik; Grundlagen der Mechanik: Statik und Dynamik 1880/81 Winter. Deutsches Museum, NL 174/450 [Mach 1880].

———. 1882. Notizbuch April 1882. Deutsches Museum, NL 174/523 [Mach 1882].

———. 1883. *Die Mechanik in Ihrer Entwickelung Historisch-Kritisch Dargestellt*. Leipzig: F.A. Brockhaus.

———. 1897. *Contributions to the Analysis of the Sensations*. Chicago: Open Court. Trans C.M. Williams. First edition, *Beiträge zur Analyse der Empfindungen*. Jena: Fischer, 1886.

———. 1911. *History and Root of the Principle of the Conservation of Energy*. Chicago/London: The Open Court Publishing Company/Kegan Paul, Trench, Trübner & Co. Trans Philip E.B. Jourdain. First edition, *Die Geschichte und die Wurzel des Satzes von der Erhaltung der Arbeit*. Prag: J.G. Calve, 1872..

Pojman, Paul. 2011a. Ernst Mach. In *The Stanford Encyclopedia of Philosophy (Winter 2011 Edition)*, ed. Edward N. Zalta, first published 2008.

———. 2011b. The Influence of Biology and Psychology on Physics: Ernst Mach Revisited. *Perspectives on Science* 19(2): 121–135.

Ratliff, Floyd. 1965. *Mach Bands: Quantitative Studies on Neural Networks in the Retina*. San Francisco/London: Holden-Day.

Staley, Richard. 2013. Ernst Mach on Bodies and Buckets. *Physics Today* 66 (12): 42–47.

Chapter 7
Scientific Inference and the Earth's Interior: Dorothy Wrinch and Harold Jeffreys at Cambridge

Teru Miyake

7.1 Introduction

That philosophical issues can, at times, have a profound influence on the development of a science, is by now a familiar idea to historically-minded philosophers of science. Studies of such influences have been limited, however, to a few sciences, with by far the most work being done on physics. I am quite confident, for example, that only a handful of philosophers are at all aware of a connection between the development of the field of seismology in the early decades of the twentieth century and the school of philosophy centered around Bertrand Russell and G. E. Moore at Cambridge. The main aim of this paper is to bring this connection to light through an examination of the work of Dorothy Wrinch and Harold Jeffreys, each of whom were students of this tradition of philosophy, and who went on to do important work in several different fields of science.

Since this is a topic that is unfamiliar to most philosophers, and the motivations for this research might be obscure, I will explain my reasons for undertaking this research. I was originally led to this work through an interest in the epistemology of contemporary geophysics. Put simply, I was interested in the question of how it is that we can say we know things about the deep interior of the earth, in the face of what looks to be a severe underdetermination problem. Almost all the information we have about the earth's deep interior comes from observations of seismic waves at the earth's surface. The number of such observations we can make is necessarily finite, and indeed limited by the number and magnitude of the earthquakes that occur. On the other hand, the earth's interior is immense, and likely to be extremely complicated—the number of degrees of freedom of the earth's interior is practically infinite. This underdetermination was explicitly pointed out in the late 1960s by the geophysicists George Backus and Freeman Gilbert (Backus and Gilbert 1968), and

T. Miyake (✉)
Nanyang Technological University, Singapore, Singapore
e-mail: tmiyake@ntu.edu.sg

F. Stadler (ed.), *Integrated History and Philosophy of Science*, Vienna Circle Institute Yearbook 20, DOI 10.1007/978-3-319-53258-5_7

this realization led to the development of techniques for making inferences in the face of uncertainty that are now known as the theory of inverse problems (Miyake 2011).

Of course, it is not as if the problem was unknown to geophysicists until then. Indeed, it is in grappling with this very problem that Harold Jeffreys, one of the founders of the field of seismology, came to work on the foundations of probability, eventually writing two seminal works on the topic (Jeffreys 1931, 1939). Of particular interest for philosophers is a series of papers that Jeffreys co-authored in the late 1910s and 1920s with Dorothy Wrinch, a philosopher and mathematician who would later go on to make significant contributions to mathematical biology. At the time, she was a student of Russell's, and an active member of philosophical circles at Cambridge. This is, then, an interesting case where budding scientists who were deeply familiar with the cutting edge of epistemology tried to take these philosophical ideas and apply them towards their own scientific work, the most significant being the work that would occupy Jeffreys for the rest of his life—that of inferring properties of the deep interior of the earth from seismological data. It turns out that Wrinch and Jeffreys found the then-current theories of epistemology lacking, and tried to develop their own ideas—some of which would end up in Jeffreys's later work on the foundations of probability.

This paper will be a brief examination of this work by Wrinch and Jeffreys, focusing on the following question: Exactly what was it that they found inadequate about the epistemological theories of the time, and how did they try to overcome such problems? In the rest of this paper, I will first give very short biographical sketches of Wrinch and Jeffreys, since many readers are likely unfamiliar with them. I will then provide a couple of examples of Wrinch and Jeffreys's confrontations with ideas coming out of Cambridge philosophy at the time, and the way in which they reacted to them.

7.2 Dorothy Wrinch and Harold Jeffreys: Biographical Sketches

Harold Jeffreys (1891–1989) is a towering figure in geophysics, widely recognized as one of the founding fathers of this field. In the 1930s and 1940s, he and his student Keith Bullen developed some of the earliest detailed earth models based on observations of travel times of seismic waves (see Bullen 1975). His textbook on geophysics, *The Earth*, went through six editions in five decades and was a standard reference for geophysicists into the late twentieth century. Jeffreys is also known to philosophers of probability for his objective Bayesian views, propounded in two books, *Scientific Inference* (1931) and *The Theory of Probability* (1939). There is substantive work by historians and philosophers on Jeffreys's later views on probability and his clashes with R. A. Fisher over the interpretation of probability (see, e.g., Howie 2002), but this paper will focus on an earlier period in his life. Jeffreys attended Cambridge in the early 1910s, and then became a fellow of St. John's College in 1914. This paper focuses on the period from the 1910s through the 1920s, when Jeffreys wrote the series of papers with Dorothy Wrinch.

Dorothy Wrinch (1894–1976) is a lesser-known figure in comparison to Harold Jeffreys, but she made important contributions to a wide range of fields, including mathematics and biology, not to mention philosophy, and has recently been the subject of a biography (Senechal 2012). Wrinch started out reading mathematics at Girton College, Cambridge, but became interested in epistemology and logic, probably through her mentor at Girton, Constance Jones, an early female logician. Wrinch was among a group of young philosophers that Bertrand Russell invited to study privately with him in London in 1916 when his lectureship at Trinity College was revoked for his outspoken opposition to the war. For the next couple of years, she studied philosophy and became a part of social circles surrounding Russell in London. Because Russell did not have an official lectureship at Cambridge anymore, Wrinch went back to Cambridge to study mathematics at the post-graduate level under the mathematician G. H. Hardy, but with Russell as a sort of background supervisor.

At some point in around 1916–1917, she befriended Harold Jeffreys, who at the time was apparently a regular attendee of Cambridge philosophy events. They discovered that they had a shared interest in the application of philosophy to science, and spent the summer of 1919 discussing the scientific method, and probably working on their first co-authored papers. This is the year of Eddington's famous solar eclipse expedition to confirm Einstein's theory of gravity, and Cambridge was buzzing with discussions of the theory of relativity and the scientific method. In the late 1910s through the 1920s, Wrinch was an active member on the Cambridge philosophical scene, publishing a dozen papers (not counting her co-publications with Jeffreys) in *Mind* and *Proceedings of the Aristotelian Society* on a variety of philosophical topics, mainly focusing on epistemology and scientific method. Later in the 1930s, she became a member of the Theoretical Biology Club, a circle of biologists including Joseph Woodger, J. D. Bernal, Joseph Needham, and Conrad Waddington, who were interested in applying the tools of logic and philosophy to problems in biology.[1] Wrinch then went on to do important work in biology, particularly on protein structure (see Senechal 2012).

7.3 The Wrinch-Jeffreys Papers

This paper focuses on an earlier period in Wrinch's life, the period in the late 1910s through the 1920s during which she and Jeffreys collaborated on a series of papers. I am aware of seven Wrinch-Jeffreys papers, published in *Nature*, *Philosophical*

[1] Daniel Nicholson and Richard Gawne gave a talk on Woodger at the Vienna conference. An interesting question which I am not at present able to answer is to what extent the philosophical views of Woodger and Wrinch were influenced by each other.

Magazine, and the *Monthly Notices of the Royal Astronomical Society Geophysical Supplement*. A quick look at their titles will give a good sense of the variety of the work they were doing[2]: "On certain aspects of the theory of probability" (1919), "The relation between geometry and Einstein's theory of gravitation" (1921), "The variable depth of earthquake foci" (1922), "On certain fundamental principles of scientific inquiry" (1921, 1923), "The theory of mensuration" (1923), and "On the seismic waves from the Oppau Explosion of 1921 Sept 21" (1923). The papers on seismology seem, on the face of it, to have very little to do with the other papers, which are on epistemology and methodology. In fact, if you read the seismology papers, you will find very little hint of any connection to the theoretical papers. But it's clear that there is a connection: Wrinch and Jeffreys were attempting to do scientific work, and at the same time they were looking for ways to provide epistemological foundations for what they were doing.

Jeffreys makes a particularly clear statement of what he was trying to do during this period in a later review he wrote of L. J. Savage's *Foundations of Statistical Inference*:

> I first took an interest in problems of scientific inference because my work concerned geophysics and cosmogony. It involved laws derived from observations on a laboratory scale and over intervals of time ranging from minutes to 2000 years, which were applied to distances of over 6000 km for the center of the Earth and 10^9 km for the solar system, and to times over 10^9 years. A usual criticism was that such extrapolation was speculative and unverifiable. The regular textbooks provided no definite answer, and the only hint was that the laws were certain in the same sense as Euclidean geometry before 1840, that no alternative was imaginable. I was unable to believe either that the laws were certain independent of experience or that extrapolation was wholly unjustified. (Jeffreys 1963, p. 407)

Consider the central problem of seismology. The problem is to determine the properties of whatever is deep within the earth's interior, 6000 km down, based on observations of seismic waves at seismographic receiving stations on the earth's surface. Jeffreys thought that it ought to be possible to arrive at knowledge about the earth's deep interior, but that the theories of epistemology of science of the day were inadequate to show how you could have such knowledge.

In their 1921 paper, "On Certain Fundamental Principles of Scientific Inquiry", Wrinch and Jeffreys start off by stating a couple of conditions that must be fulfilled by any acceptable theory of scientific method:

> In order that a scientific method may be of any value, it must satisfy two conditions. In the first place, it must be possible to apply it in the actual cases to which it is meant to be relevant. In the second, its arguments must be sound. The main object of science is to increase knowledge of the world, and if a method is not applicable to anything in the world it obviously cannot lead to any knowledge. This principle is very elementary, and it is probably for that very reason that it is habitually overlooked in theories of scientific knowledge. (Wrinch and Jeffreys 1921, p. 369)

Wrinch and Jeffreys believed that there were no existing theories of the epistemology of science that could fulfill the two requirements of applicability and soundness

---

[2] For the specific references for all of the papers, see the Bibliography in Jeffreys (1971).

simultaneously—in particular, that "in all cases it is found that they fail to satisfy the criterion of applicability in practice" (p. 369). But exactly what were the existing theories of scientific knowledge that they were objecting to?

7.4 Wrinch and Jeffreys on Russell's Epistemology of Science

In the following paragraphs, Wrinch and Jeffreys describe a tension between philosophers and scientists who write on the epistemology of science. The fundamental problem is that existing theories of epistemology make it extremely difficult, or even impossible, for us to have scientific knowledge. Philosophers generally have regarded inductive generalization as formally fallacious, and this casts doubt on the possibility of scientific knowledge. Scientists, on the other hand, start with the belief that their methods work (the results speak for themselves!), and most of them would condemn philosophy before admitting any epistemic shortcomings. Wrinch and Jeffreys then mention more recent writers who have tried to resolve this problem by showing that scientific knowledge is possible without the use of inductive generalization, by the addition of certain fundamental postulates. Unfortunately, they say, these attempts are unsatisfactory because there is no reason to think that such postulates are true.

Here, Wrinch and Jeffreys explicitly name Bertrand Russell and A. N. Whitehead as such writers, and they say that their objection to them is fundamentally the same—that these epistemological views require an uncritical use of postulates that involve infinite classes of entities. Here, I will focus on their rejection of Russell's theory, for which they cite the essay "The Relation of Sense-Data to Physics", published in *Mysticism and Logic* (Russell 1917). Here is what Russell has to say about the epistemology of physics:

> In physics as commonly set forth, sense-data appear as functions of physical objects: when such-and-such waves impinge upon the eye, we see such-and-such colors, and so on. But the waves are in fact inferred from the colors, not vice-versa. Physics cannot be regarded as validly based upon empirical data until the waves have been expressed as functions of the colors and other sense data.
>
> Thus if physics is to be verifiable we are faced with the following problem: Physics exhibits sense-data as functions of physical objects, but verification is only possible if physical objects can be exhibited as functions of sense-data. We have therefore to solve the equations giving sense-data in terms of physical objects, so as to make them instead give physical objects in terms of sense-data. (Russell 1917, pp. 146–147)

This is a statement of the epistemology of physics as involving a kind of *inverse problem*. That is, Russell is saying that you can think of physics as providing a function that maps physical objects onto sense-data—given some particular layout of physical objects and a human, physics provides a way of calculating, in principle, what sense-data that human would experience. But the epistemology of physics is the opposite of this. You start with sense-data, and you are supposed to arrive at physical objects. Russell states this process as taking the functions from physical objects to sense-data, and inverting them.

It's worth pointing out here the parallel between the problem stated by Russell, and a problem that Harold Jeffreys was really interested in. Think of the problem of determining the properties of the deep interior of the earth based on observations of seismic waves at its surface. Given the density and the values of the elastic constants at each point in the earth's interior, you can uniquely determine what observations you ought to have at the earth's surface, such as the travel times of seismic waves. Physics provides us with a function that maps the properties of the earth's interior to seismic wave observations. But the problem that Jeffreys was interested in is the inverse of this problem—given the data we record at the surface, what are the density and values of the elastic constants at each point in the earth's interior? This is, in fact, exactly the problem that Jeffreys attempts to solve later in the 1930s and 1940s—the determination of detailed models of the earth's interior based on travel times of seismic waves. The problem, as Jeffreys realized very well, is that the inverse problem is *massively* underdetermined—in fact, radically different earth models could be compatible with observations at the earth's surface. And this is still, I think, *the* central problem in the epistemology of seismology.

But let's get back to Russell. How does Russell propose to solve this inverse problem? Here is Russell's solution:

> The supreme maxim in scientific philosophizing is this: *Wherever possible, logical constructions are to be substituted for inferred entities.* [...] This method, so fruitful in the philosophy of mathematics, will be found equally applicable in the philosophy of physics, where I do not doubt it would have been applied long ago but for the fact that all who have studied this subject hitherto have been completely ignorant of mathematical logic. (Russell 1917, pp. 155–157, emphasis in the original)

An example that Russell gives is irrational numbers. Instead of *inferring* the existence of irrational numbers, you *define* irrational numbers in terms of a cut between two sets of rational numbers. This gets rid of the problematic inference to mysterious entities that cannot be accessed directly, namely irrational numbers. The proposal is to construct all of physics out of sense-data in a similar manner, getting rid of problematic inferences to inaccessible entities. And this is supposed to solve the problem of inversion that Russell states.

Now, whatever the merits of this solution in mathematics, Wrinch and Jeffreys believe that it will not work for physics. The approach that Russell takes towards solving the inverse problem takes the existence of physical objects to be epistemologically problematic, and he suggests the construction strategy to avoid this existence problem. But the problem in which Wrinch and Jeffreys are interested, while having a similar structure, is entirely different. This becomes clear when you think about trying to apply this procedure to an actual science. Suppose you want to determine the properties of the deep interior of the earth, based on observations at its surface, as indeed Wrinch and Jeffreys wanted. The *existence* of things in the earth's deep interior is not at all in question. What they want to determine are the *properties* of these things—for example, what is the value of the density 1000 km from the earth's center?

Now, suppose you tried to define values for the properties of the physical objects as constructions out of sense-data. A feature of Russell's picture is that he proposes to define physical objects not just as constructions out of sense-data, but constructions out of *sensibilia*—things that are like sense-data, but which are not actually experienced by anyone. On this picture, physical objects are then infinite classes of sensibilia. But in order to define specific values for the properties of the physical objects, you would have to guarantee that these infinite classes would converge on certain limits. Wrinch and Jeffreys point out that there is no such guarantee. Now, you could try to choose values for the unobserved sensibilia in such a way that the infinite classes would converge. But this then would be arbitrary—many different sets of values for the properties of the sensibilia could be compatible with any given set of values for the sense-data. The sense-data would underdetermine the properties of the physical objects.

Wrinch and Jeffreys do not state this explicitly, but as I just mentioned, the real problem with the inversion procedure, is not *existence*, but *nonuniqueness* (or underdetermination). And this is clear if we go back to the case of determining properties of the earth's interior based on observations at its surface. Given any set of observations at the Earth's surface, more than one distribution of properties in the Earth's interior could be compatible with those observations, and some of them may be radically different from each other.

7.5 Wrinch and Jeffreys on a Probabilistic Epistemology of Science

Wrinch and Jeffreys thus conclude that construction is an unfeasible strategy, and that any epistemology of science would have to include principles of inductive inference. Jeffreys has stated (Jeffreys 1963, p. 407) that he was influenced as a student by Karl Pearson's *The Grammar of Science* (1911). Pearson has an explicitly probabilistic approach to establishing scientific facts, according to which we can infer facts about physical objects based on observations. This is much preferable to Russell's approach, from the point of view of Wrinch and Jeffreys. But there was a serious problem with probabilistic approaches such as Pearson's at the time.[3] In order to establish scientific facts, Pearson suggests the use of Laplace's Rule of Succession. This is the famous rule through which Laplace calculated what the probability that the sun would rise tomorrow would be, given that it has risen every day for the past 5000 years. The rule states that if an event has occurred m times in succession, then the probability that it will occur again is given by $(m + 1)/(m + 2)$.

In 1918, C. D. Broad, another young Cambridge philosopher familiar to Wrinch and Jeffreys, published a paper in which he does a re-analysis of the Rule of

[3] In addition to the objection I describe here, they reject Pearson's frequentist interpretation of probability for a logicist one, but debates between frequentists and logicists have been covered in detail elsewhere (e.g, Galavotti 2005, Howie 2002), so I will not examine this here.

Succession that would be suitable for cases of enumerative induction of the familiar "all ravens are black" type (Broad 1918). For example, suppose the total number of ravens is n. And suppose you have observed m ravens, and they have all turned out to be black. Broad shows that the probability that all ravens are black would then be given by $(m + 1)/(n + 1)$. This means, of course, that you would have to observe a large proportion of all existing ravens before you could have any confidence that all ravens are black. But it seems reasonable to make the inference much earlier than that. Wrinch and Jeffreys thus take the C. D. Broad result to show that Laplace's theory of probability is inadequate for an epistemology of science, because it would make knowledge of such universal generalizations much more difficult than seems reasonable.

Now, it's pretty easy to see what the problem is. When carrying out the calculation, Broad makes the initial assumption that all possible proportions of black ravens are equally probable—each possible proportion is in effect an alternative hypothesis, all of which are initially taken to be equally probable. He then uses Bayes' theorem to calculate the posterior probabilities after having observed m black ravens. Thus, the prior probability, before any observations are made, that all the ravens are black, is $1/n$, where n is the total number of ravens. If n is large, then the hypothesis that all the ravens are black gets assigned a very low prior probability. Moreover, because all of the hypotheses are assigned equal prior probabilities, all of the hypotheses that have yet to be eliminated as possibilities (that is, all hypotheses for which the number of black ravens is equal to or greater than m) are assigned equal posterior probabilities. According to this picture, there is no way to adjust the relative probabilities between these various hypotheses, even though it seems intuitively obvious that we should ultimately assign a high probability to one particular hypothesis—namely, the one according to which all the ravens are black. The obvious solution, which Wrinch and Jeffreys (1919) suggest, is to initially weight the priors so that some hypotheses have a higher initial probability than the others. In the case of the ravens, we would want to assign much higher initial priors to the hypothesis that all ravens are black.

Wrinch and Jeffreys later (1921) describe a more general version of this idea, which they call the "Simplicity Postulate". They start off with an example problem, that of determining the law governing the motion of an object rolling down an inclined plane. Suppose you are given the position of the object at five different times during its descent, so you have five data points. They point out that an infinite number of laws that agree with these observations can be found. So how do you choose among these laws? The naïve answer here is that you would find the simplest law that describes these motions. But how do you determine the simplest law, and what justification can you give for making this choice? More generally, suppose we think of science as, essentially, a procedure in which we take observations and try to find laws that agree with those observations. For any given number of observations, we will always find an infinite number of laws that agree with those observations. How do we choose among these laws?

Wrinch and Jeffreys believe that the correct approach is a probabilistic one along the lines of Pearson's, but the problem is that we would not want to assign equal priors to all of the infinite number of possible laws. For one thing, since there are an infinite number of laws, if you assigned non-zero priors to all the laws, it seems that they would sum to infinity. For another, it seems obvious that you would want to weight the simpler laws more. Wrinch and Jeffreys come up with a solution, but one that is likely to be looked upon with extreme skepticism by most philosophers. They first ask us to consider the fact that you can have converging infinite sums such as 1/2 + 1/4 + 1/8 + ... which will sum to unity. There are, of course, many other examples of such converging infinite sums. So Wrinch and Jeffreys decide that you can assign priors to the infinite number of possible laws on two conditions: (1) the number of possible laws is countable (e.g., aleph-null), and (2) the possible laws can be put into an order that corresponds to their simplicity.

The first condition is hard to argue for, since the number of possible functions is clearly uncountable, even if you restrict the set of functions to analytic functions. Further, if there are numerical constants in the laws, and these constants can vary continuously, the number of possible values of the constants would also be uncountable. But Wrinch and Jeffreys argue that we can appropriately restrict the set of possible laws so that they obey the following principle:

> Every quantitative law can be expressed as a differential equation of finite order and degree, in which the numerical coefficients are integers. (Wrinch and Jeffreys 1921, p. 386)

It would be beyond the scope of this paper to examine in detail the arguments that Wrinch and Jeffreys offer for this principle.[4] But if this can be done, then the number of quantitative laws is countable. And they then suggest a way of putting these laws in order of simplicity. First, you group equations into those that have equal values of the sum of the order, degree, and absolute values of the coefficients. Each such group has a finite number of members. You then go through and arrange each member of this group in an appropriate way.[5] At the end of the procedure, the equations occurring earlier in the series will have low order and degree, and the numerical coefficients will be small integers. The equations that occur earlier in the series will be assigned higher prior probabilities, and although the number of equations is infinite, the probabilities can still be assigned in such a way that they will sum to unity.

I suspect that most philosophers, upon hearing of this solution, will find this solution to be extremely *ad hoc*.[6] How can you simply declare, for example, that probabilities can be appropriately assigned so that they will sum to unity? Wrinch and Jeffreys themselves were surely aware that philosophers would be highly critical of their solution, but they do little more than argue that such an assignment of probabilities should be *possible*, and appear to have no direct argument that this is how the probabilities *ought* to be assigned. Jeffreys (1931), when discussing the simplicity postulate, gives an indirect argument for this method—that it would, among other things, allow us to

---

[4] It is given in Wrinch and Jeffreys (1921), pp. 383–386.

[5] The procedure is described in detail in Jeffreys (1931), p. 46.

[6] But see Braithwaite (1931) for an insightful, critical examination of Jeffreys's method.

> …discard at first sight the suggestion that the perihelion of Mercury could be explained if the attraction of the sun varied inversely as the 2.000000016 power of the distance instead of the exact inverse square.[7] The exiguous prior probability of such a law puts it beyond consideration, apart from the inconsistency with the observed motion of the moon's perigee that led to its abandonment. In fact the law established with a high probability by experience is not an approximation to the simple law, but the exact simple law itself. Consequently extrapolations over an indefinitely wide range can be carried out with the full probability of the law. This is the justification of the inferences concerning conditions at the center of the earth or millions of years ago that form so large a part of geophysics and cosmogony. (Jeffreys 1931, pp. 50–51)

This should not, I believe, be read as some kind of loose transcendental argument for the simplicity postulate. Rather, this goes back to the original motivations that Wrinch and Jeffreys had for examining the epistemology of science. They wanted an epistemology that could be applied to actual sciences such as seismology, and an epistemology according to which scientific knowledge is impossible just could not do the job.

7.6 Conclusion

Wrinch and Jeffreys are not philosophers, and that is the reason they end up with an epistemology of science that would be unacceptable to most philosophers. In saying this, I do not mean that they were ignorant of philosophy. On the contrary, their deep knowledge of the cutting edge of epistemology of the time is apparent both through their writings, and the things that have been written about them by contemporary philosophers such as C. D. Broad and R. B. Braithwaite. The difference is in the motivation. Wrinch and Jeffreys needed an epistemology of science that could be applied to actual sciences, such as seismology, and the epistemology they came up with, though flawed, was one that they believed could be so applied, something that they thought could not be said for any of the leading theories of epistemology of the time.

I said that my original motivation for studying the work of Wrinch and Jeffreys is an interest in the epistemology of modern seismology. Wrinch and Jeffreys believed that the only way out of a massive underdetermination problem, such as the one faced by seismologists, is through something like the simplicity postulate. The problem is that the simplicity postulate seems to be unjustified—the only argument for it seems to be that scientists do it, and it seems to work generally. My own view is that Wrinch and Jeffreys made a mistake by turning to epistemology to attempt to justify the simplicity postulate—it might be better justified by appealing to methodology, not epistemology.[8] But that would be the subject for another paper. In any case, an examination of the later works of Wrinch and Jeffreys, in which they lay out

[7] This is a reference to the American astronomer Asaph Hall, who made this suggestion (see Jeffreys 1931, p. 186).

[8] See Miyake (2013) for a more detailed exposition of this idea.

a theory of "mensuration", as well as the further development of the epistemological views of Jeffreys in response to the needs arising out of his work in seismology, would be of further philosophical interest.

References

Backus, G.E., and J.F. Gilbert. 1968. Numerical Applications of a Formalism for Geophysical Inverse Problems. *Geophysical Journal of the Royal Astronomical Society* 13: 247–276.

Braithwaite, R.B. 1931. Scientific Inference by Harold Jeffreys; Science and First Principles by F. S. C. Northrop; Critique of Physics by L. L. Whyte. *Mind, New Series* 40 (160): 492–501.

Broad, C.D. 1918. On the Relation Between Induction and Probability (Part I.). *Mind, New Series* 27 (108): 389–404.

Bullen, K.E. 1975. *The Earth's Density*. London: Chapman and Hall.

Galavotti, Maria Carla. 2005. *Philosophical Introduction to Probability*. Stanford: CSLI Publications.

Howie, David. 2002. *Interpreting Probability: Controversies and Developments in the Early Twentieth Century*. Cambridge: Cambridge University Press.

Jeffreys, Harold. 1931. *Scientific Inference*. Cambridge: Cambridge University Press.

———. 1939. *Theory of Probability*. Oxford: Oxford University Press.

———. 1963. The Foundations of Statistical Inference by L. J. Savage. *Technometrics* 5 (3): 407–410.

———. 1971. *Collected Papers of Harold Jeffreys on Geophysics and Other Sciences: Vol. 1, Theoretical and Observational Seismology*. London: Gordon and Breach.

Miyake, Teru. 2011. *Underdetermination and Indirect Measurement*. Dissertation submitted to Stanford University.

———. 2013. Essay Review: Isaac Newton's Scientific Method. *Philosophy of Science* 80 (2): 310–316.

Pearson, Karl. 1911. *The Grammar of Science, Third Edition*. London: Adam and Charles Black.

Russell, Bertrand. 1917. *Mysticism and Logic, Second Edition*. London: George Allen and Unwin Ltd.

Senechal, Marjorie. 2012. *I Died for Beauty: Dorothy Wrinch and the Cultures of Science*. Oxford: Oxford University Press.

Wrinch, Dorothy, and Harold Jeffreys. 1919. On Some Aspects of the Theory of Probability. *Philosophical Magazine, Sixth Series* 38: 715–731.

———. 1921. On Certain Fundamental Principles of Scientific Inquiry. *Philosophical Magazine, Sixth Series* 42 (249): 369–390.

Chapter 8
Values, Facts and Methodologies. A Case Study in Philosophy of Economics

Thomas Uebel

8.1 Introduction

One question that has long haunted philosophy of science is whether facts and values are so inextricably mixed up in social science that objectivity in any sense robust enough to distinguish its findings from mere opinion becomes unattainable. A not uncommon view nowadays is that such entanglement only shows the untenability of conceptions of objectivity that forbid it and that a new and value-sensitive conception of objectivity needs to be developed. While the discussion in recent years has centred on the issue of how estimations of inductive risk incur judgments of value[1]—and so generalize the issue across all of the sciences—it is worthwhile to remember that in the decades around the previous turn of the century when the social sciences became established as such, it was the more or less direct interference of politics in the process of social scientific fact finding that was the focus of concern and prompted the demand for value-neutrality. This older worry has not, I submit, lost its urgency and it may be salutary to consider whether, especially in sciences issuing in policy advice, value entanglement is inevitable there as well. I will present a case study from what may at first appear most hostile territory, namely one of the most value-laden of all areas in the social sciences, the socialist calculation debate in political economy.[2]

At issue throughout the socialist calculation debate was a viability claim, namely whether it was possible to put into practice the principles of a socialist economy.

[1] See, e.g., Douglas (2000) and (2009) elaborating an argument from Rudner (1953).

[2] While I am not unsympathetic to the argument that it is not social science but social philosophy, I shall here treat political economy as a representative of normatively exposed social science and the episode in question as offering a particularly sharp challenge to the general demand for value-neutrality in social science.

T. Uebel (✉)
Department of Philosophy, University of Manchester, Manchester, M13 9PL, UK
e-mail: thomas.uebel@manchester.ac.uk

F. Stadler (ed.), *Integrated History and Philosophy of Science*, Vienna Circle Institute Yearbook 20, DOI 10.1007/978-3-319-53258-5_8

What was in contention in the episode of that debate that is in focus here was the methodology used to achieve a putative impossibility result. Its proponent claimed that his impossibility result was dismissed on account of a purely politically motivated methodological decision against it. The threat accordingly would be this: methodologies that either allow or disallow politically sensitive impossibility results are themselves subject to the suspicion of having been chosen with political ends in mind or, less conspiratorial, of inherent bias. As will be shown, however, a far more persuasive successor to the impossibility argument in question does not depend on the contentious methodology—though it reduced the modality in question back to empirical implausibility. This suggests that the allegation that opposition to the method of argument originally used was wholly political was mistaken and that the fact-method-value entanglement did not extend as far as was feared.

8.2 The Socialist Calculation Debate and Its Background: The *Methodenstreit*

Ever since Albert Schäffle's *Die Quintessenz des Sozialismus* of 1875, theorists of socialism were confronted by the challenge to explain how the coordination of the supply of and demand for goods and commodities was to be achieved by economic planning when the labor theory of value seemed patently unable to draw the distinctions required for this and no replacement was in sight.[3] Prominent Marxists like Karl Kautsky tended retain the labor theory of value despite his and, later, Eugen von Böhn-Bawerk's criticisms.[4] A different response by Otto Neurath gained currency in 1919: that was to abandon the labor theory of value as well and argue for an economy in-kind, for marketless socialism.[5] Noting that the institution of a market meant the implementation of decision procedures that allocated resources irrespective of the satisfaction of social needs, Neurath argued that a truly socialist economy had to replace monetary calculation for profit wholesale by calculation in-kind in light of these social needs. It was against this that Ludwig von Mises claimed the impossibility result that concerns us here.[6]

---

[3] For Schäffle, supply and demand under socialism "would fall into a hopeless quantitative and qualitative discrepancy" (1875/1892, 87). Schäffle also raised the other two standard objections to socialism: how its institution could overcome the presumed problem of resultant motivation deficits on part of the workers, and how the infringement of the sovereignity of choice of labor and consumption was to be handled.

[4] See Kautsky (1902) and Böhm-Bawerk (1896).

[5] For examples of Neurath's socialisation plans, see his (1919) or (1920a) and (1920b).

[6] This was first published in Mises (1920) and then greatly expanded into Mises (1922). The ongoing discussions were commented on in Mises (1924) and (1928).

The rarely noted background of this particular episode of the socialist calculation debate at issue here (it continue to last much longer)[7] was the so-called *Methodenstreit*, the dispute about the proper method to be employed in the study of economics, between Carl Menger, founder of the Austrian School of economists, and Gustav Schmoller, then heading the German Historical School.

Originated in 1883–84 and lasting in the main until around the turn of the century, the *Methodenstreit* resists simple categorization even in strictly methodological terms.[8] For instance, Menger's subjectivist theory of value did not develop in isolation from mainstream German economics and Schmoller's *Verein für Sozialpolitik* had many Austrian members.[9] And likewise, Schmoller was not a naive inductivist and Menger not an uncompromisingly radical a priorist deductivist. To be sure, they differed in method, but the dispute is better characterized by the opposition of an early hypothetico-deductive empiricism (Schmoller) and a tempered essentialism that granted the need for testing applied theories even though the theoretical components themselves were a priori (Menger).[10] And while, notoriously, both parties alleged that the other slavishly and quite unjustifiably followed the methods of natural science, again the reality was different. Schmoller admired Wilhelm Dilthey's goal to develop the "true methodology of *Geisteswissenschaft*", while Menger affirmed that in the realm of human phenomena exact laws can be established under the same formal presupposition as in the realm of natural phenomena.[11]

What the debate did do was throw into relief the different foci of the participants. Historical economics was concerned with individual actions and/or institutions and sought non-strict generalizations on basis of complete descriptions, while theoretical economics was concerned with economic action in a generic sense and sought strict laws with help of the "composite method" (one form of methodological individualism under another name). Moreover, the different parties also had different policy orientations. While the theorists of the Historical School advocated state involvement in social reform, the Austrians typically supported a strictly *laissez-faire* approach to economic policy.[12] In consequence, Schmoller missed in Menger a macroeconomic perspective to hang a reform programme on, while Menger aimed for theory of spontaneous evolution of social institutions through convergence of individual economic action that rendered intervention unnecessary.

---

[7] Not all treatments of the socialist calculation debate cover it in its entirety; for two that do see Hutchison (1953, Ch. 18) and Steele (1992, Ch. 4).

[8] For the origin see Menger (1883/1963), Schmoller (1883/1888) and Menger (1884/1970). For summaries of the *Methodenstreit* see Ringer (1968, Ch.3, Sect. 2) or Hands (2001, 72–94).

[9] See Streissler (1990a) and Grimmer-Solem (2003, 721), respectively.

[10] See Hansen (1968, 146–151) and Smith (1990).

[11] See Schmoller (1883/1888, 304) and Menger (1883/1985, Appdx V, Fn)

[12] Needless to say, there were exceptions to this generalization, but see, with regards to Schmoller, Hansen (1968, 158), Nau (2000, 508), Grimmer-Solem (2003,252), and with regard to Menger, Streissler (1990b).

In sum: even though both of their positions were far more nuanced than their opponents admitted, there remained plenty to disagree about by the opponents in the *Methodenstreit*, albeit along unpredictable battle lines.

8.3 The *Methodenstreit* Restarted

After the initial hostilities between Schools the debate was allowed to fizzle out on both sides after the turn of the century. Yet that was not a correct outcome according to the *magnum opus* of the doyen of later-generation Austrian economists, Mises.

> In the *Methodenstreit* between the Austrian economists and the Prussian Historical School ... much more was at stake than what kind of procedure was the most fruitful one. The real issue was the epistemological foundations of the science of human action and its logical legitimacy. Starting from an epistemological system to which praxeological thinking was strange and from a logic which acknowledged as scientific—besides logic and mathematics—only the empirical natural sciences and history, many authors tried to deny the value and usefulness of economic theory. Historicism aimed at replacing it by economic history; positivism recommended the substitution of an imaginary social science which should adopt the logical structure and pattern of Newtonian mechanics. Both these schools agreed in a radical rejection of all the achievements of economic thought.[13]

Never mind that Mises arrogated to himself and his school the exclusive right to reprsent "economic thought". Note first that, for him, the Historical School denied the possibility of universal economics altogether whereas the Austrians only denied that universal laws be furnished by the historical sciences. Yet more than mere theory had been in play according to an earlier assessment of his:

> The Historical School in Europe and the Institutionalist School in America are harbingers of the ruinous economic policy that has brought the world to its present condition and will undoubtedly destroy modern culture if it prevails.[14]

For Mises not just error or plain stupidity prompted the Historical School's opposition to the solution to the *Methodenstreit* proposed by him in the tradition of the Austrian School. Rather, he wrote, "political, not scientific, considerations are decisive".[15]

So, for Mises, the *Methodenstreit* (i) was far from over, (ii) involved all of social science, (iii) concerned the proper understanding of abstract theory and its priority over history, and (iv) assumed a political dimension. For Mises, the *Methodenstreit* extended into the socialist calculation debate. Let's now consider Mises' economic methodology more closely.

---

[13] Mises (1940/1949, 4).

[14] Mises (1933b/1960, xvii.)

[15] Ibid.

8.4 Economics as an *a priori* Science

Mises rejected (Millian) empiricism (and chided Menger for undue concessions). Instead Mises adopted von Wieser's idea that economics builds on

> a fund of experiences that are the common possession of all who practice economy. These are the experiences that every theorist already finds within himself without first having to resort to special scientific procedures. They are the experiences concerning facts of the external world, as for instance, the existence of goods and their orders; experiences concerning facts of an internal character, such as the existence of human needs, and concerning the consequences of this fact; and experiences concerning the origin and course of economic action on the part of most men.[16]

Note, however, that this "common experience" was to be "sharply distinguished" (far more than Wieser did) from "experience in the sense of the empirical sciences", indeed, it is "the very opposite of it". Economics builds on "that which logically precedes experience and is, indeed, a condition and presupposition of every experience": it builds on *a priori* elements.[17]

Mises thought this apriority to be comparable to that of logic and mathematics:

> The science of human action that strives for universally valid knowledge is the theoretical system whose hitherto best elaborated branch is economics. In all of its branches, this science is *a priori*, not empirical. Like logic and mathematics, it is not derived from experience; it is prior to experience. It is, as it were, the logic of action and deed.[18]

The logic of economic action was *a priori* due to the conceptual necessities it followed and expressed. The justification of its claims did not recur to sense experience.

> What we know about the fundamental categories of action—action, economizing, preferring, the relationship of means and ends, and everything else that, together with these, constitutes the system of human action—is not derived from experience. We conceive all this from within, just as we conceive logical and mathematical truths a priori, without reference to any experience. Nor could experience ever lead anyone to the knowledge of these things if he did not comprehend them from within himself.[19]

The apriority of economics thus consisted in this.

> As thinking and acting men, we grasp the concept of action. In grasping this concept we simultaneously grasp the closely correlated concepts of value, wealth, exchange, price, and cost. They are all necessarily implied in the concept of action, and together with them the concepts of valuing, scale of value and importance, scarcity and abundance, advantage and disadvantage, success, profit, loss. The logical unfolding of all these concepts and categories in systematic derivation from the fundamental category of action and the demonstration of the necessary relations among them constitutes the first task of our science. The part that deals with the elementary theory of value and price serves as the starting point in its

[16] Mises (1933a/1960, 21).

[17] Ibid., 22.

[18] Ibid., 12–13.

[19] Ibid., 13–14.

> exposition. There can be no doubt whatever concerning the *a priorist* character of these disciplines.[20]

Needless to say, this leaves important questions open about the nature of the *a priori* that Mises invoked, which cannot be pursued here. Suffice it to say, however, that with his insistence that the necessities discerned the facts of the matter, Mises seems closer to Menger's Aristotelian essentialism than Kant's transcendental idealism. In any case, however, it is important to note that these methodological views were not late accretions to Mises' doctrines (though their elaboration was). Already in the "Introduction" to the first German edition of *Socialism* in 1922, Mises declared that economics was part of *Geisteswissenschaft*:

> Social science … finds its objects within, not in the external world. … In this sense, too, the question must be answered—no longer of importance today—whether social science belongs to natural science or *Geisteswissenschaft*. Social life is part of … *Geist*.[21]

With his reference to *Geisteswissenschaft*. of course, Mises raised the question of just how he conceived of its difference from natural science.

Importantly, while affirming economics as a *Geisteswissenschaft*, Mises rejected existing conceptions of a distinct "*Verstehen*", like Dilthey's, as precluding the possibility of objective knowledge. He also rejected Max Weber's methodology for interpretive sociology.

> The laws of sociology are neither ideal types nor average types. Rather, they are the expression of what is to be singled out of the fullness and diversity of phenomena from the point of view of the science that aims at the cognition of what is essential and necessary in every instance of human action. … The causal propositions of sociology … express that which necessarily must always happen as far as the conditions they assume are given.[22]

Mises's science of human action, "praxeology", was precisely such *a priori* sociology, a science of the essence of human action.

What is most importantly in this connection is that Mises also rejected Max Weber's conception of different types of rationality (substantive, instrumental, affective and traditional).[23] Thus he wrote that "everything that we regard as human action … is instrumentally rational: it chooses between given possibilities in order to attain the most ardently desired goal".[24] For Mises, all rationality was instrumental in character, so economic rationality was elevated to rationality *per se*.

---

[20] Ibid., 24.

[21] Mises (1922, 11–12, trans. TU). The sections §§ 4–5 of the Introduction from which this quotation is taken were dropped from the second German edition and so were never made it into the translation.

[22] Ibid., 90–91.

[23] See Weber (1921/1978, 24–26).

[24] Mises (1929/1960, 85, trans. Restored from "rational" to "instrumentally rational").

8.5 Mises' Argument Against Marketless Socialism

Let's now turn to what many consider Mises' master argument against socialism.[25] Here we must first note the specificity of its target. It was Neurath's conception of marketless socialism that Mises targeted in passages like these.

> Calculation *in natura*, in an economy without exchange, can embrace consumption-goods only; it completely fails when it comes to deal with goods of a higher order. And as soon as one gives up the conception of freely established monetary price for goods of a higher order, rational production becomes completely impossible. Every step that takes us away from private ownership of the means of production and from the use of money also takes us away from rational economics.[26]

To be sure, Mises conceded (as did Weber around the same time)[27]:

> In the narrow confines of a closed household economy, it is possible throughout to review the process of production from beginning to end, and to judge all the time whether one or another mode of procedure yields more consumable goods. This, however, is no longer possible in the incomparably more involved circumstances of our own social economy.[28]

What Mises stressed was that the rational allocation of resources in complex economies requires the value of production goods to be calculated—which presupposes a market with commensurate monetary exchange values.

> Without economic calculation there can be no economy. Hence in a socialist state wherein the pursuit of economic calculation is impossible, there can be—in our sense of the term—no economy whatever. In trivial and secondary matters rational conduct might still be possible, but in general it would be impossible to speak of rational production any more. There would be no means of determining what was rational, and hence it is obvious that production could never be directed by economic considerations.[29]

In sum: "any economic system of calculation would become absolutely impossible ... There is only groping in the dark. Socialism is the abolition of rational economy."[30] A socialist economy is a rationalist impossibility.

8.6 Mises' Characterization of His Opposition

Mises was not shy about mentioning the opposition by name in his polemics against marketless socialism from the years 1920, 1922 and 1928, but not in 1933 and 1940 when he put the spotlight on the complicity of historicist thinking in what he regarded as elementary mistakes in political economy. But it was not long before

[25] This was first published in Mises (1920) and then greatly expanded into Mises (1922).

[26] Mises (1920/1935, 104).

[27] See Weber (1921/1978, 103)

[28] Mises (1920/1935, 103).

[29] Ibid., 105.

[30] Ibid., 109–110.

Mises returned to form, albeit now recognizing his old and presumably vanquished foe re-emergent in a new guise. Thus Mises wrote about the proposal for marketless socialism in 1949:

> It would hardly be worthwhile even to mention this suggestion if it were not the solution that emanated from the very busy and obtrusive circle of the 'logical positivists' who flagrantly advertise their program of the 'unity of science'. Cf. The writings of the late chief organizer of this group, Otto Neurath, who in 1919 acted as the head of the socialization bureau of the short-lived Soviet republic of Munich, especially his [1919], pp. 216 ff.[31]

When Mises wrote this, Neurath had died but the philosophical movement with which he was associated was reaching its apex of influence in the country where Mises was now in exile. There Mises' argumentation turned to epistemology and the attack on his own methodological separatism of *Geisteswissenschaft*. Neurath served him well again and his "guilt" was transferred to the movement. Thus Mises wrote:

> The most obtrusive champion of the neopositivist program concerning the sciences of human action was Otto Neurath who, in 1919, was one of the outstanding leaders of the short-lived Soviet regime in Munich and later cooperated briefly in Moscow with the bureaucracy of the Bolsheviks. *Knowing they cannot advance any tenable argument against the economists' critique of their plans, these passionate communists try to discredit economics wholesale on epistemological grounds.*[32]

And still in 1962 Mises issued this indictment:

> The way in which the philosophy of logical positivism depicts the universe is defective. It comprehends only what can be recognized by the experimental methods of the natural sciences. It ignores the human mind as well as human action. … The socialist or communist prepossession and activities of outstanding champions of logical positivism and 'unified science' are well known. … Otto Neurath instilled into the methodological monism of 'unified science' its definite anticapitalistic note and converted neopositivism into an auxiliary of socialism and communism. Today both doctrines, Marxian polylogism and positivism, amicably vie with each other in lending theoretical support to the 'Left'.[33]

Note first what Mises alleged here: that methodological arguments were used by his opponents to support an unsustainable theoretical position purely for political ends: methodology was subordinated to political maneuvering. Note second that this criticism of logical positivism runs squarely counter to and positively contradicts the refrain of denouncements of logical positivism's supposedly unreflective technocratic conformism emanating from Max Horkheimer and his collaborators and successors in the Frankfurt School.[34] This suggests that both criticisms be assessed carefully, but here we stick to Mises.

---

[31] Mises (1949, 699).

[32] Mises (1957, 242, emphasis added).

[33] Mises (1962/2002, 113, 116, 119).

[34] See Horkheimer (1937); for discussion see O'Neill and Uebel (2004).

8.7 Neurath's Opposition to Mises' Argument

Neurath was unimpressed by Mises' putative impossibility result. On the one hand, he sought to defend his socialisation plans as a plausible alternative—albeit one that was dependent on success in the research programme of "universal statistics" so as to allow the "calculation in kind" of need satisfaction to be properly developed that socialist resource allocation depended on.[35] On the other hand, Neurath revived his ecological argument which claims the impossibility of accounting objectively for sustainability considerations by monetary calculation.[36]

Of these two strategies the second strikes me as more promising but, unfortunately, it did not gain the prominence at the time which it deserves. Neurath's reason for holding it to be impossible to account objectively for sustainability considerations by monetary calculation was that the monetarisation of environmental goods and particularly of the discount rate for the disutility suffered by future generations due to the exhausted resources was arbitrary.[37] Moreover, to begin with, sustainability considerations and their like required awareness of the incommensurability of the values involved; once that was recognized, the relevant decisions had to be arrived at by democratic deliberation—precisely the kind of process that was pre-empted by the automaticism of the profit motive in the market economy.

So Neurath challenged Mises' conception of rationality—importantly, however, he did not challenge its possibility, but only his monopolistic claim for it. Is it true, he asked, that, as Mises put it, "we are unable to grasp the concept of economic action and of economy without implying in our thought the concept of economic quantity relations and the concept of an economic good"?[38] Neurath could grant this (at least for the sake of the argument at issue), but he pressed further. Does rationality demand that all these quantity relations must be expressed in cardinal measures? On this point Neurath disagreed ever since 1911 when he started developing algebaic calculi for representing goods transfers using only comparative measures: "a unitary measure is not a necessary condition for comparability"[39] (It was, of course, precisely these merely comparative calculi that evaluations of aggregate allocations of resources individually assessed by calculations in-kind were to depend on.) Taking account of the incommensurability of values meant refraining from the commensuration by monetarisation that the market required.

Now Neurath did not discuss Mises' *Geisteswissenschaft* methodology explicitly. It does not follow, however, that Mises' claim to have faced methodological opposition was wrong. What prompted Neurath's disregard of Mises' impossibility

[35] For the research program-defense see Neurath (1925a/2004, 446).

[36] See Neurath and Schumann (1919, 15–16), Neurath (1925b/2004, 468–471), Neurath (1928/1973, 263); for discussion see Uebel (2005).

[37] This is a point shared by contemporary ecological economists: see Martinez-Alier (1995).

[38] Mises (1933a/1960, 14).

[39] Neurath (1911/1998, 473). Here lies the origin of his later "Inventories of Standards of Living" (1937).

result was, after all, dissent from his methodological starting point. Mises claimed a conceptual impossibility for marketless socialism. Neurath traced this claim back to Mises' *a priorist* presuppositions about rational agents and the nature of economic rationality. Neurath rejected these presuppositions. He proposed to deal with recalcitrant socio-economic phenomena (e.g. to effect transparent sustainability calculations) on anti-apriorist methodological grounds: social science had to be done empirically, not by pressing phenomena into a scheme of intuited essences. While allowing for variation between the disciplines, Neurath's program of unified science prescribed an empirical methodology for both the natural and the social sciences and rejected the separatism of *Geisteswissenschaft* on epistemological grounds: scientific reasoning required evidence that was intersubjectively accessible—which intuitions of essences are not. That calculation in kind did not issue in the determination of an optimum solution via commensuration of heterogeneous values did not disqualify it from the start.

So Mises was right that methodological considerations played into the socialist calculation debate between himself and Neurath, but he was wrong to claim that these considerations were only a cover for politically motivated dissent. There are good reasons to keep social science empirical: not only Neurathian socialists will want to dispute the adequacy of the methodology behind Mises' argument.[40]

8.8 A Different Anti-socialist Calculation Argument

As my intention here is not to defend Neurath's marketless socialism, let me turn to the next stage of this aspect of the socialist calculation debate to reinforce my methodological point.

From about 1935 onward, Friedrich von Hayek developed Mises' calculation argument into something more readily appreciable. There has been dispute between Austrian economists as to the degree of independence of Hayek's development of Mises' argument.[41] Suffice it to say here that the grounds of Mises' impossibility claim lie not in practical limitations but the principled absence of a market which precludes monetary calculation of resource allocation. Soon, of course, the dialectic of the calculation debate was changing in that marketkless socialism was no longer championed by advocates of planned economies; instead different forms of market socialism were being discussed according to which the coordinating task of the market was taken over by planners who mimicked the market in preparatory calculations before issuing directives for the command economy.[42] Nevertheless, just as

[40] Note that even a sympathetic critic like Robert Nozick, who is open to the possibility of synthetic necessary truths, cast doubt on Mises's conception of the status of his theory of action (1977, 361–369).

[41] The seed of Hayek's information argument can be found at Mises (1920/1935, 102), but

[42] See, e.g., Lange (1936).

Mises' argumentation was readily extended to this, so Hayek's development of it remained applicable in full to Neurath's proposal.

Let's consider three characteristic formulations of Hayek's argument:

> In a centrally planned society the selection of the most appropriate among the known technical methods will be possible only if all that knowledge can be used in the calculations of the central authority. This means in practice that this knowledge will have to be concentrated in the heads of one or at best a very few people who actually formulate the equations to be worked out. It is hardly necessary to emphasize that this is an absurd idea even in so far as that knowledge is concerned which can properly be said to 'exist' at any one time.[43]

Here note that Hayek's argument shifts the focus from pressing home the irrationality of supposing that resource allocations can be determined independently of a market and its monetary measures to conveying the enormity of practical demands that the institution of a planned economy would impose. The point is repeated in the form of a positive evaluation of the market alternative some 5 years later:

> … main merit of real competition is that through it use is made of knowledge divided by between many persons which, if it were to be used in a centrally directed economy, would all have enter the single plan. To assume that all this knowledge would be automatically in the possession of the planning authority seems to me to miss the main point.[44]

And another 5 years later Hayek summed up the point as follows:

> To assume all the knowledge to be given to a single mind in the same manner in which we assume it to be given to us as the explaining economists is to assume the problem away and to disregard everything that is important and significant in the real world.[45]

It may be noted that, over time, talk of "absurdity" gave way to talk of "missing the point" and ignoring what's practical in the "real world". In line with this, Hayek's argument against socialism can be seen to ring significant changes on the Misean blueprint. To begin with, his argument concerns not what is rational *per se* in economic matters, but whether all the required information can be gathered. Moreover, his "information argument" also can be seen to shift (i) the methodological ground and (ii) the status of the claim at issue. No longer is it necessary to presuppose an *a priori* methodology to assent to the conclusion. (Whether Hayek himself intended this is another question but that very many of those who accept the argument do not accept Misean apriorism is surely significant.) With Hayek, (i) the epistemological ground of the argument against (marketless) socialism is shifted away from Mises' *Wesensschau* to empirical considerations. Moreover, (ii) Mises' categorical impossibility claim is dropped for plausibility considerations.

Needless to say, even a *prima facie* evaluation of Hayek's argument force against (marketless) socialism is beyond the remit of this paper.[46] What is important for present purposes is this methodological point. Unlike Mises' rationality-centered version, Hayek's information version of the anti-socalist calculation argument can

[43] Hayek (1935a, b)/1948, 155).

[44] Hayek (1940/1948, 202).

[45] Hayek (1945/1948, 91).

[46] For a recent critical discussion see O'Neill (2012).

be read to proceed on grounds that are at least in principle acceptable to empiricists, even to theorists beholden to some version of the unity of science thesis. (Whether Neurath did in fact accept it, and whether Hayek's argument is taken to hold against all forms of socialism, not only marketless socialism, are matters not at issue here.)

8.9 Conclusion

Our case study concerned a perceived threat to the probity of social science. If the result in a given field has policy implications, does it follow that the methodological choices made in the course of reaching it reflect and are determined by political choices? Since our case concerned political economy, it may have looked like it does indeed follow, but, despite one protagonist's claim that this holds for his opponents, it does not. As we saw, there were—and are—good reasons to oppose Mises' methodology.

It may be wondered whether all supposedly inextricable entanglements of facts and/or methodologies on one side and moral and/or political values on the other, can be so disentangled. Nothing I showed here suggests this—nor was it meant to do so. My point has been rather to warn against generalizations and to suggest that each case of entanglement has to be considered on its own merit. What the case at hand would seem to show is that even in political economy it is sometimes possible to rebut a claim to the end of objectivity on account of value-entanglement—still before a wholesale reconceptualization of objectivity is attempted—and to uphold, at least in that instance, the goal of value-neutrality.[47]

References

Böhm-Bawerk, Eugen von. 1896. Zum Abschluss des Marxschen Systems. In *Staatswissenschaftliche Arbeiten. Festgabe für Karl Knies*, ed. O.V. Boenigk, Berlin. Trans. *Karl Marx and the Close of his System*, Kelley, New York, 1949.

Douglas, Heather. 2000. Inductive Risk and Values in Science. *Philosophy of Science* 67: 559–579.

———. 2009. *Science, Policy and the Value-Free Ideal*. Pittsburgh: University of Pittsburgh Press.

Grimmer-Solem, Eric. 2003. *The Rise of Historical Economics and Social Reform in Germany 1864–1894*. Oxford: Clarendon Press.

Hands, Wade. 2001. *Reflection Without Rules. Economic Methodology and Contemporary Science Theory*. Cambridge: Cambridge University Press.

Hansen, Reginald. 1968. Der Methodenstreit in den Sozialwissenschaften zwischenGustav Schmoller und Karl Menger. In *Beitraege zur Entwicklung der Wissenschaftstheorie im 19. Jahrhundert*, ed. A. Diemer, 137–173. Meisenheim: Hain.

[47] To be sure, this argument also presupposes the distinction between epistemic and non-epistemic values. Steel (2010) has provided a solid defence against recent challenges to that distinction.

Hayek, Friedrich von. 1935a. [Socialist Calculation II:] The Present State of the Debate. In Hayek 1935b, 201–243. Reprinted in Hayek 1948, 148–180.
———, ed. 1935b. *Collectivist Economic Planning Critical Studies of the Possibility of Socialism*. London: Routledge.
———. 1940. [Socialist Calculation III:] The Competitive Solution, *Economica* 7. Reprinted in Hayek 1948, 181–208.
———. 1945. The Use of Knowledge in Society. *American Economic Review* 35: 519–530. Reprinted in Hayek 1948, 77–91.
———. 1948. *Individualism and Economic Order*. Chicago: University of Chicago Press.
Horkheimer, Max. 1937. Der neueste Angriff auf die Metaphysik. *Zeitschrift für Sozialforschung* 6: 4–51. Trans. "The Latest Attack on Metaphysics" in Horkheimer, *Critical Theory. Selected Essays*. New York: Continuum, 1972.
Hutchison, Terence. 1953. *A Review of Economic Doctrines, 1870–1929*. Oxford: Clarendon Press.
Kautsky, Karl. 1902. *Die soziale Revolution. Bd. 2: Am Tage nach der Revolution*. Dietz, Stuttgart, Trans. *The Social Revolution and On the Morrow of the Revolution,* London, 1907.
Lange, Oskar. 1936. "On the Economic Theory of Socialism", Review of Economic Studies 4, reprinted. with additions and changes in B. E. Lippincott (ed.), *On the Economic Theory of Socialism*, University of Minnesota Press, Minneapolis, 1938, reprinted McGraw-Hill, New York, 1964, 55–143.
Martinez-Alier, Juan. 1995. Political Ecology. Distributional Conflicts and Economic Incommensurability. *New Left Review* 21: 70–88.
Menger, Carl. 1883. *Untersuchungen ueber die Methode der Socialwissenschaften und der Politischen Oekonomie insbesondere*, Vienna. Trans. *Problems of Economics and Sociology*, University of Illinois Press, Urbana, 1963. Repr. as *In vestigations into the Method of the Social Sciences with Special Reference to Economics*. New York: New York University Press, 1985.
———. 1884. *Die Irrtuemer des Historismus in der deutschen Nationaloekonomie*. Vienna. Reprinted in Menger, *Gesammelte Werke. Band III* (ed. F.A. Hayek), 2nd edition, Mohr, Tuebingen, 1970, 1–98.
Mises, Ludwig von. 1920. Die Wirtschaftsrechnung im sozialistischen Gemeinwesen. *Archiv für Sozialwissenschaft und Sozialpolitik* 47: 86–121. Trans. "Economic Calculation in the Socialist Commonwealth" in Hayek 1935b, 78–130.
———. 1922. Die *Gemeinwirtschaft. Untersuchungen ueber den Sozialismus*. Fischer, Jena. 2nd enlarged ed. 1932. Trans. *Socialism. An Economic and Sociological Analysis*. London: Cape, 1936, 2nd enlarged ed. 1951.
———. 1924. Neue Beiträge zum Problem der sozialistischen Wirtschaftsrechnung. *Archiv für Sozialwissenschaft und Sozialpolitik* 51: 488–500.
———. 1928. Neue Schriften zum Problem der sozialistischen Wirtschaftsrechnung. *Archiv für Sozialwissenschaft und Sozialpolitik* 60: 187–190.
———. 1929. Soziologie und Geschichte. Epilog zum Methodenstreit in der Nationaloekonomie. *Archiv für Sozialwissenschaft und Sozialpolitik* 61: 465–512. Reprinted in Mises 1933b. Trans. "Sociology and History", in Mises 1960, 68–129.
———. 1933a. Aufgabe und Umfang der allgemeinen Wissenschaft vom menschlichen Handeln. In Mises 1933b. Trans. "The Task and Scope of the Science of Human Action", in Mises 1960, 1–67.
———. 1933b. *Grundprobleme der Nationaloekonomie. Untersuchungen ueber Verfahren, Aufgaben und Inhalt der Wirtschafts- und Gesellschaftslehre*, Jena.
———. 1940. *Nationaloekonomie. Theorie des Handelns und Wirtschaftens*. Editions Union, Geneva. Reprinted Philosophia, Munich, 1980.
———. 1949. *Human Action. A Treatise on Economics*. London: Hodge and Company.
———. 1957. *Theory and History. An Interpretation of Evolution and History*. New Haven: Yale University Press.
———. 1960. *Epistemological Problems of Economics*. Princeton: Van Nostrand.

———. 1962. *The Ultimate Foundation of Economic Science. An Essay on Method.* Van Nostrand, Princeton. 2nd edition reprinted Liberty Fund, Indianapolis, 2002.

Nau, Heino Heinrich. 2000. Gustav Schmoller's Historico-Ethical Political Economy: Ethics, Politics and Economics in the Younger German Historical School, 1860–1917. *European Journal of the History of Economic Thought* 7: 507–531.

Neurath, Otto. 1911. Nationalökonomie und Wertlehre, eine systematische Untersuchung. *Zeitschrift für Volkswirtschaft, Sozialpolitik und Verwaltung* 20: 52–114, reprinted in Neurath, *Gesammelte ökonomische, soziologische und sozialpolitische Schriften I* (ed. R. Haller and Ulf Höfer), Hölder-Pichler-Tempsky, Vienna, 1998: 470–518.

———. 1919. Wesen und Weg der Sozialisierung. In Neurath, *Durch die Kriegwirtschaft zur Naturalwirtschaft*, Callwey, Munich, 1919, 209–220. Trans. "Character and Course of Socialisation" in Neurath 1973: 135–150.

———, 1920a, Ein System der Sozialisierung.,*Archiv für Sozialwissenschaft und Sozialpolitik* 48, trans. A System of Socialisation" in Neurath 2004: 345–370.

———. 1920b. *Vollsozialisierung*. Diederichs, Jena. Trans. "Total Socialization" in Neurath 2004: 371–404.

———. 1925a. *Wirtschaftsplan und Naturalrechnung*. Laub, Berlin. Trans. Economic Plan and Calculation in Kind in Neurath 2004: 405–465.

———, 1925b, Sozialistische Nützlichkeitsrechung und kapitalistische Reingewinnrechnung. *Der Kampf* 18, trans. Socialist Utility Calculation and Capitalist Profit Calculation, in Neurath 2004: 466–472.

———, 1928, *Lebensgestaltung und Klassenkampf*, Laub, Berlin. Trans. Personal Life and Class Struggle in Neurath 1973: 249–298

———. 1937. Inventory of the Standard of Living. *Zeitschrift für Sozialforschung* 6: 140–151. Reprinted in Neurath 2004: 513–526.

———. 1973. In *Empiricism and Sociology*, ed. M. Neurtah and R.S. Cohen. Dordrecht: Reidel.

———. 2004. In *Economic Writings. Selections 1904–1945*, ed. T.E. Uebel and R.S. Cohen. Dordrecht: Kluwer.

Neurath, Otto, and Wolfgang Schumann. 1919. *Können wir heute sozialisieren?Eine Darstellung der sozialistischen Lebensordnung und ihres Werdens*. Leipzig: Klinkhardt.

Nozick, Robert. 1977. On Austrian Methodology. *Synthese* 36: 353–392.

O'Neill, John. 2012. Austrian Economics and the Limits of the Market. *Cambridge Journal of Economics* 36: 1073–1090.

O'Neill, John, and Thomas Uebel. 2004. Horkheimer and Neurath: Restarting a Disrupted Debate. *European Journal of Philosophy* 12: 75–105.

Ringer, Fritz. 1968. *The Decline of the German Mandarins, The German Academic Community 1890–1933*. Cambridge, MA: Harvard University Press.

Rudner, Richard. 1953. The Scientist qua Scientist Makes Value Judgments. *Philosophy of Science* 20: 1–6.

Schäffle, Albert. 1875. *Die Quintessenz des Sozialismus*. Perthes, Gotha. Trans. *The Quintessence of Socialism*. London: Swan Sonnenschein, 1892.

Schmoller, Gustav. 1883. Die Schriften von K. Menger und W. Dilthey zur Methodologie der Staats- und Sozialwissenschaften. *Jahrbuch fuer Gesetzgebung, Verwaltung und Volkswirtschaft im Deutschen Reich* 7: 975–994. Reprinted In Schmoller, *Zur Litteraturgeschichte der Staats- und Sozialwissenschaften*. Duncker & Humblot, Leipzig, 1888: 275–304.

Smith, Barry. 1990. Aristotle, Menger, Mises: An Essay in the Metaphysics of Economics. *In Caldwell* 1990: 263–288.

Steel, Daniel. 2010. Epistemic Values and the Argument from Inductive Risk. *Philosophy of Science* 77: 14–34.

Steele, D. 1992. *From Marx to Mises. Post-Capitalist Society and the Challenge of Economic Calculation*. La Salle: Open Court.

Streissler, Erich. 1990a. The Influence of German Economics on the Work of Menger and Marshall. *In Caldwell* 1990: 31–68.

———. 1990b. Carl Menger on Economic Policy: The Lectures to Crown Prince Rudolf. *In Caldwell* 1990: 107–131.

Uebel, Thomas. 2005. Incommensurability, Ecology and Planning. Neurath in the Socialist Calculation Debate, 1919–1928. *History of Political Economy* 37: 309–341.

Weber, Max. 1921. *Wirtschaft und Gesellschaft. Grundriss der verstehenden Soziologie*. Mohr (Siebeck), Tübingen, 4th rev. ed. by J. Winckelmann, 1956. Transl. *Economy and Society. An Outline of Interpretive Sociology* (ed. by. G. Roth and C. Wittich), Berkeley: University of California Press, 1978.

Chapter 9
Extended Evolution and the History of Knowledge

Jürgen Renn and Manfred Laubichler

9.1 Introduction

This paper provides a framework for analyzing the history of knowledge from the perspective of extended evolution, a conceptual framework that analyzes evolutionary processes as transformations of extended regulatory network structures, and is designed to apply to a whole range of phenomena, from genome and biological to cultural and technological evolution. Regulatory networks, such as gene regulatory networks or institutions, control the behavior of individual elements within systems, whether these are genes within cells or organisms or individuals within societies. All of these phenomena can be seen as a form of extended evolution. Our framework is inspired by Ernst Mach, a scientist turned historian and philosopher who developed a distinctly evolutionary conception of knowledge. For Mach the dynamics of highly structured systems of knowledge, such as science, was a logical outgrowth of the evolutionary roots of human cognition. He focused specifically on the role of memory—information from genomes to cultural traditions in present-day terminology—and emphasized how all life forms extract "knowledge" or information through a continuous process of trial and error. As a consequence of these processes of knowledge acquisition, tested "hypotheses" are incorporated into the (genetic or cultural) make-up or memory of each species. Indeed, Mach's ideas about the cultural transmission of shared or collective memories and the role of institutions in that process are an early version of what we call cultural evolution today (Mach 2011).

J. Renn (✉)
Max Planck Institute for the History of Science, Berlin, Germany
e-mail: renn@mpiwg-berlin.mpg.de

M. Laubichler
Arizona State University, Tempe, AZ, USA
e-mail: Manfred.Laubichler@asu.edu

F. Stadler (ed.), *Integrated History and Philosophy of Science*, Vienna Circle Institute Yearbook 20, DOI 10.1007/978-3-319-53258-5_9

The proposal of extended evolution is a conceptual framework for the evolution of complex systems based on the integration of regulatory network and niche construction theories (Laubichler and Renn 2015). It applies equally to cases of biological, social and cultural evolution. A general feature of this framework is the transformation of complex networks through the linked processes of externalization and internalization of causal factors between regulatory networks and their corresponding niches. Externalization refers to the stable or lasting transformation of niches (biological, cultural, social and technological) through the actions of systems, whereas internalization captures those processes that lead to the incorporation of stable features of the environment(s) into the regulatory structures governing the actions of systems. These processes extend previous evolutionary models and focus on several challenges, such as the path-dependent nature of evolutionary change, the dynamics of evolutionary innovation and the expansion of inheritance systems (Laubichler and Renn 2015). Extended Evolution is in several ways a further development from previous models of cultural evolution that have focused on extending standard evolutionary dynamics to different domains (e.g. Richerson and Boyd 2005; Tomasello 2014) and of niche construction (e.g. Laland et al. 2008). The focus of our theory of extended evolution is both on the transformations of regulatory structures and the generation of variation and novelty and the resulting dynamics of evolutionary change.

9.2 Cultural Evolution

Cultural processes do not suspend biological evolution but interact with it. The extent to which cultural processes themselves are subject to an evolutionary logic has been controversially discussed. There are in any case striking parallels between biological and cultural history. The history of human societies amounts to a hereditary process involving populations in which cultural information is transmitted with variations from one generation to the next. Human societies come into being and disappear, competing with each other for resources. In spite of these parallels, it is notoriously difficult to define what cultural information is, what counts as variation and what the units of selection in cultural evolution are (Richerson and Christiansen 2013).

Cultural evolution, on the other hand, clearly displays features that have gained prominence in recent developments of evolutionary theory, such as niche construction and the role of complex regulatory networks. Human societies transform their environments by means of their material culture which forms a "niche" and decisively shapes their evolution. Human societies do not just vary randomly, but in ways that are governed by hierarchically organized internal structures regulating the behavior of their members. From this perspective, cultural information is stored in these regulative structures as well as in the material culture that supports them. The structures govern variation and the units of selection at the hierarchical level at which they are effective.

Cultural evolution even suggests more precisely how complex regulatory networks interact with niche construction. Niches do not just affect fitness landscapes but extend the system itself by providing crucial regulatory effects. At the same time, the transformation of the environment by human societies constitutes a transformation of their informational organization. While current discussions no longer focus on reducing cultural evolution to a neo-Darwinian model that demands a particulate character or on the blindness of cultural evolution, they do emphasize classical Darwinian concepts, such as transmission, variation, selection, drift and migration (Mesoudi et al. 2011), and not the transformation of regulative structures that are otherwise central to the social sciences and cultural history. Institutions are primarily studied with regard to the behavioral mechanisms that enable them, and with regard to their consequences for group selection processes. Niche construction, on the other hand, has become increasingly relevant to discussions about cultural evolution (Jeffares 2012; Laland et al. 2000, 2008; Mace and Holden 2005; Mesoudi et al. 2004; Odling-Smee 1995; Read 2014; Richerson and Boyd 2005; Richerson and Christiansen 2013; Wimsatt 2013). Both niche construction and regulatory structures are key factors in the study of certain episodes, for example when considering the role of the domination of fire or the invention of lethal weapons for human sociality (Bowles and Gintis 2011). However, so far this has not yet led to a general approach that deals with their ongoing evolutionary interaction.

The framework of extended evolution presented here avoids ongoing discussions about the relationship between biological and cultural evolutionary processes as it focuses on the general systems' properties governing all domains of evolution. Extended evolution is therefore both an abstract formulation of evolutionary dynamics based on the transformation of regulatory structures and a series of domain specific interpretations of individual cases of evolution. Furthermore, the systems' perspective of extended evolution theory also explains how traditionally separate domains of evolution—biology, culture, knowledge—can affect each other, as can be seen in the well-known feedback between developments in knowledge and culture to biology in the context of the Neolithic Revolution. The evolution of lactose tolerance as a consequence of domestication is just one such example.

9.3 The History of Knowledge as a Case of Extended Evolution

We consider cultural evolution as a special case of extended evolution (Laubichler and Renn 2015). Here, the interaction between cognition and tradition as regulatory structures, as well as their external representations as part of the material culture form the niche that plays a key dynamical role (Damerow 1996, 2000). More precisely, from this perspective, cultural evolution deals with embedded networks of actors, actions and their results (see Fig. 9.1).

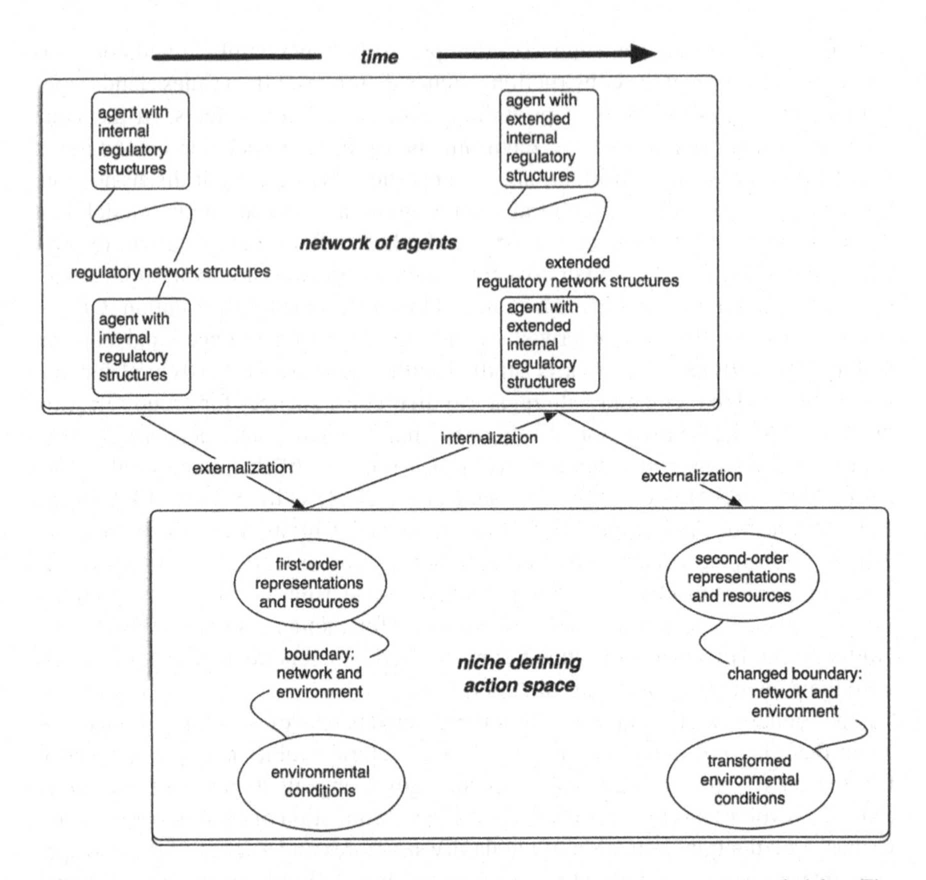

Fig. 9.1 Networks of agents evolve, including their internal regulatory structures and niches. The niche itself has a network structure induced by the primary network. Its nodes are those aspects of the environment that condition, mediate or become the target of actions, in short the environmental resources and conditions of the internal system. The extended network, including the environment, defines an action space that shapes possible innovations, canalizes the evolutionary process and delimits the structure of the inheritance system

A crucial aspect of such a network is its capability to learn: the actors are assumed to have an internal, cognitive structure that governs the coordination of their actions and that in turn can change as a result of a reflection on their actions. The encoded experience of the actors constitutes their knowledge. Actions are always embedded in a larger network that includes other actors and the environment, and also a material and social culture resulting from prior actions. Action networks can generate systemic structures that we designate as institutions, regulating the actions that preserve these structures. The material and social culture corresponds, in biological terms, to a niche that has been constructed by a species transforming its environment in such a way as to affect its own living conditions.

The material means are part of the context employed by the actor to reach the goal of an action. They comprise, in particular, the tools available to a given culture and the useful material resources found in the environment. They also constrain the range of actions that are possible in a given situation, thus defining a horizon of possibilities for actions. In dealing with challenges to the preservation of its actors and systemic structures, a network of actors can transform them into an enrichment and reorganization of its regulative structures. Such innovations are possible because the horizon of possibilities inherent in a given material context is larger than anticipated by any given set of actors, a principle that may be considered as a non-teleological version of Hegel's cunning of reason or a phylogenetic version of Vygotsky's "zone of proximal development" (Damerow 1996; Lock 2000; Vygotsky 1978).

Knowledge and institutions are two important regulative structures that govern actions within such a network (Renn 2014, 2015a). We will first discuss knowledge, then institutions. Knowledge is, as mentioned above, encoded experience. Based on experience, it is, at the same time, the capacity of an individual, a group or a society to solve problems and to anticipate appropriate actions. In short, knowledge is a problem-solving potential. But it is not just a mental structure. It also involves material and social dimensions that play a crucial role in determining which actions are possible and legitimate in a given historical situation. Knowledge may be shared within a group or a society. Material artifacts such as instruments or texts may be used in learning processes organized by societal institutions, allowing individuals to appropriate the shared knowledge.

The social and material dimensions of knowledge are therefore critical for understanding its transmission from one generation to the next. We designate the societal structures governing its production, dissemination and appropriation as the knowledge economy of a society. For most of human history, this knowledge economy was not supported by specialized institutions such as schools or universities. Knowledge was rather transmitted from generation to generation as part of a society's self-reproduction by raising children and involving them in labor processes and various cultural activities (Renn 2012).

Institutions are here defined as a means of reproducing the social relations existing within a given group or society and, in particular, collaborative roles and the societal distribution of labor. The coordination of individual actions mediated by institutions often presupposes behavioral norms and belief systems such as habits, religion, law, morality or ideology. Institutions represent the potential of a society or a group to coordinate the actions of individuals and to interact with their environment. In the most general sense, institutions can be conceived as encoded, collective experience, which results in sets of shared behaviors connected by cognitive, social and material links.

With their "potential for action," institutions therefore bear close relations to knowledge as we have defined it, but there are also important differences. While there is no knowledge without the mental anticipation of actions, institutions must largely regulate cooperative behavior without such direct mental anticipation of collective actions and their consequences. Nevertheless, institutions involve knowledge on various levels and must embody and transmit this knowledge in the sense of the capacity of individuals to anticipate actions that are compatible with the coordination regulated by institutions. In addition, institutions must also transmit knowledge on social control and on how to resolve conflicts. And finally, institutions form the basis for the knowledge economy of a society. The history of knowledge must therefore be studied in close conjunction with the history of institutions in this broader sense.

All contexts of action may serve as an external representation of the two key regulative structures we have been considering: knowledge and institutions. Such external representations can be used to share and transmit but also to transform these regulative structures. We have already seen that external representations such as artifacts, tools and texts play a key role in the societal transmission of knowledge; they may also serve as the means of actions performed in order to process and transform knowledge, such as tables or computers.

As for institutions, all kinds of material aspects—tools, machines, persons, symbols or rituals—can become part of their external, material representation. They then represent a normative social order, defining a field of actions that is compatible with the regulations of an institution. The coordination of individual interactions can be partly transfered onto the external representations of an institution, such as following a command chain, dealing with paperwork in an administration, exchanging goods for money on the market, or applying written law to a violation of norms.

We thus recognize two essential, complementary features of the model of cultural evolution we are proposing: the role of complex regulative network structures such as knowledge and institutions, and niche construction such as the creation and transmission of a material culture that includes the external representations of these regulative structures. The crucial point now for understanding the evolutionary dynamics of this system is the fact that this niche construction not only depends on complex regulative structures but also in turn shapes them. In the following, we will consider some important turning points in human cultural evolution to assess the current discussions on these turning points from the vantage point of our framework.

9.4 From Biological to Cultural Evolution

The conceptual framework of extended evolution is general enough not to force upon us a premature distinction between biological and cultural evolution, nor to reduce the latter to a metaphorical generalization of the former. An embedded network of actors, actions and their results may describe a population of our prehuman ancestors just as well as they describe a human society. The relevant regulative structures and environments will be different, but the concept is wide enough to allow for the identification of processes connecting one with the other in an evolutionary continuum.

The evolutionary mechanism giving rise to specifically human ways of thinking is often described in terms of distinct thresholds, involving ecological circumstances that drive humans into more cooperative ways of life and foster adaptations for dealing with problems of social coordination. In his recent book *A Natural History of Human Thinking* (2014) Michael Tomasello emphasizes the cooperative nature of human thinking. He postulates two key evolutionary steps. In the first step, a novel type of small-scale collaboration in human foraging led to socially shared joint goals and joint attention, creating a possibility for individual roles and perspectives within *ad hoc* situations. In the second step, which is characterized by growing human populations competing with each other, humans developed collective intentionality, enabling them to construct a common cultural ground by means of shared cultural conventions, norms, and institutions. The evolutionary mechanism is described in terms of ecological circumstances driving humans into more cooperative ways of life and fostering adaptations for dealing with problems of social coordination.

Against the background of extensive empirical studies involving comparisons between children and apes, Tomasello identifies the specific cognitive abilities emerging in these two evolutionary steps, which are designated as joint and collective intentionality, respectively. Joint intentionality is characterized by the fact that humans can conceptualize the same situation under different perspectives, that they can make recursive inferences about each other's intentional states and that they can evaluate their own thinking with respect to the normative perspectives of others. Collective intentionality extends joint intentionality to include a conventional dimension of these cognitive capabilities which are now broadly shared within a culture and no longer a matter of *ad hoc* situations.

In this picture, specific forms of human thinking thus become first the presupposition and then the consequence of human culture. Less emphasis, on the other hand, is given to the evolving results of human thinking in the form of knowledge, institutions and material culture emerging and accumulating from human interactions with their environment over time. Instead of postulating two distinct evolutionary steps leading from biological to cultural evolution, our model suggests a continuously working feedback mechanism in which the ecological circumstances cited by Tomasello as evolutionary driving forces are themselves partly created by the regulative structures of human evolution through niche construction. Therefore, we would like to emphasize, more than Tomasello did, the material aspect of human actions, not only their instrumental but also their representational aspects, which are crucial for the transmission and transformation of the evolving regulative structures. Material representations of thinking, for instance, may function as external memory, as catalysts for the emergence of different perspectives and as triggers for reflection, and thus affect all the dimensions of thinking processes mentioned by Tomasello.

9.5 The Evolution of Language

The difference between an assumption of such distinctive evolutionary steps and the perspective of extended evolution becomes strikingly clear when we turn to our first example, the problem of the origin of language and more generally of communicative structures among actors. These structures arise in collaborative situations and depend on the specific constellation, size and ensuing challenges of the relevant communities. Even before the first proto-linguistic communication systems came into being, there must have existed some regulative patterns of cooperation, such as situative action coordination mediated by visual and other material clues. At least 1.8 million years ago, at the time of *Homo habilis*, these regulative structures had already been shaped by a shared material culture of tool use and transmission. Communication systems, including gestures, pointing, facial expressions, pantomiming and, much later, vocalizations, initially would have only marginally supported such regulative structures without representing the full range of cooperative possibilities (Dediu and Levinson 2013). Rather, they may have begun as sporadic, domain-specific and highly context-dependent communicative interactions, which complemented other regulative structures and inherited their "meaning" from these structures.

While communication systems presuppose certain cognitive capabilities on the side of the actors, such as joint attention, they also affect the development of these capabilities by opening up an explorative space in the Vygotskyan sense discussed above (Lock 2000; Damerow 2000). This space exists precisely because communication systems constitute, like the underlying material culture, external representations of regulative structures that typically have a larger horizon of applicability than that given by their initial purpose or circumstances of application. This opens up the possibility for an iterative process of language evolution, resulting in the layered structure of the human communication system that we actually observe today, which comprises gestures, facial expressions, pointing, pantomiming and vocalizations, all as part of one integrated multi-modal system of human communication. In a path-breaking paper, Levinson and Holler (2014) have suggested that this multi-modal system is indeed the result of a superposition of evolutionary layers. But what kind of evolution could produce such a layered structure?

Tomasello refers to the famous treatment of *major transitions* in evolution by Maynard Smith and Szathmáry (1995), pointing out that "humans have created genuine evolutionary novelties via new forms of cooperation, supported and extended by new forms of communication. ... And humans have done this twice, the second step building on the first" (Tomasello 2014, 141). For Maynard Smith and Szathmáry, the major transitions involve changes in the way information is stored and processed, listing as the three main possibilities duplication and divergence, symbiosis and epigenesis. Their own evolutionary account of language, for example, involves the explanation of the genesis of grammatical structures by genetic assimilation, essentially turning cultural into biological inheritance.

Such an argument may fit well with the concept of human cognitive evolution being characterized by major evolutionary transitions, as is assumed by Tomasello (2014, 127) who claims that language "plays its role only fairly late in the process. ... Language is the capstone of uniquely human cognition and thinking, not its foundation." Dediu and Levinson (2013), on the other hand, argue "that recognizably modern language is likely an ancient feature of our genus pre-dating at least the common ancestor of modern humans and Neanderthals about half a million years ago." According to their view, human evolution is a more protracted and reticulated process, involving both vertical and horizontal processes of gene-cultural coevolution and leading to the multi-layered regulatory structure of the human communication system observed today.

In our view, it is precisely the combination of regulatory networks and niche construction that may account for this structure. The evolution of human communication systems is thus governed by the same dynamics that is at the forefront of an emerging new synthesis in evolutionary theory. The process must have started from some contingent ecological context that constituted an external scaffolding for human social interactions, such as conditions favoring collective foraging or hunting. These initially fragile social interactions must have involved some context-dependent signaling, for instance, by means of gestures mimicking actions. The next step would then be a gradual exploration and extension of this situative action coordination, including a discovery of new possibilities such as the ritualization and conventionalization of gestures. The point is that this exploration effectively changed the environment in which social interactions took place, creating a new niche with feedback on the action coordination itself, in particular, on the possibilities for anticipating, by means of communicative acts, the articulation of goals of actions and hence the separation of their planning and execution, as well as the division of labor.

As a consequence, actions were then performed in a new context, accompanied by an ever more extended communication system. As we have emphasized, this extension of the communication system enriched the possibilities of action coordination. The enhanced coordination may also be considered as an internalization of the contingent external conditions of social interaction-which are now turned into intrinsic properties of this interaction. In short, what may have been initially sporadic, situation-dependent signals within an originally only marginal communication process were eventually transformed into elements of a more and more self-sustaining system of communication, comprising, for instance, conventionalized gestures that are used outside of immediate action contexts. These elements hence receive their meaning not only from the contexts of action in which they are being applied but also from their role in the emerging communicative system. One immediate effect of such an internalization of external contexts is the stabilization of the originally fragile social scaffolding that may have been highly dependent on contingent external factors, for instance, specific ecological conditions.

A further consequence of the exploration of the inherent potential of an incipient communication system is the bootstrapping of developmental possibilities that can only be exploited under appropriate external conditions. This is what is designated in psychology as Vygotsky's zone of proximal development, that is, the difference between what an actor can do spontaneously without help and what the actor can do with support from a favorable environment (Vygotsky 1978; Smith et al. 1997). This difference is not only familiar from children but also from acculturated apes brought into a human environment. In our case, however, the environment favoring ontogenetic development must, of course, be itself constructed in the course of historical evolution as a cultural niche. In agreement with arguments by Lock and Damerow (Lock 2000; Damerow 2000), we claim that in the evolution of populations, systems of external representation provide the conditions for the elaboration of implications that correspond functionally to Vygotsky's zone of proximal development in individual learning.

Following Peter Damerow (1996), we further assume that on the level of the individual, action coordination is ruled by cognitive structures that are built up in ontogenesis through interaction with the environment and, furthermore, that these cognitive structures are shaped by the material means of actions. Indeed, the material means available in a given situation determine not only a horizon of possibilities for actions but also what is generic about an action, that is, what can be transferred from one context to another, thus shaping the cognitive structures emerging from their usage. Under this assumption, the construction of a cultural niche encompassing both material means for interacting with the environment and an external system of communication represented by bodily signals must in turn affect the cognitive structures acquired by individuals during their development.

This then is the beginning of a further step in the bootstrapping process: the relation between material means and external representations of thinking and communication, on the one hand, and cognitive structures on the other is, of course, not one-sided, with the actions forming the basis and cognition following suit. Newly developed cognitive structures may in turn enable an improvement of the material means of actions employed; similarly these cognitive structures can be represented in communication processes by new kinds of external representations, thus giving rise to an iterative, albeit highly path-dependent evolutionary process that generates cognitive structures of ever greater generality, in the sense of an increasing decontextualization.

It is quite conceivable that such an iterative evolution can indeed account for the emergence of the multimodal system of human communication of which modern language forms the capstone. It is, in any case, a characteristic feature of the process we describe here that new layers do not replace earlier ones but are rather integrated with the pre-existing layers in an ever more extended regulative architecture. Consider the fact, for instance, that our vocal language continues to be accompanied by body language. But rather than illustrating this in detail for the example of language, we now turn to much more recent periods of history to demonstrate the generality of our model of cultural evolution.

9.6 The Neolithic Revolution

Our second example deals with the so-called Neolithic Revolution (Renn 2015a, b). Just as there were probably many pathways leading to early communication systems, there were certainly also many routes to food production in different parts of the world. Here we will concentrate on the emergence of food production in the Fertile Crescent around 10,000 BC. Developed agriculture is a comprehensive subsistence strategy involving intensive human labor. It represents an economic system by which human societies produce a large part of their food and other conveniences from domesticated plants and animals. Domesticated plants such as cereals are adapted to human nutritional needs and even rely on human intervention for their reproduction.

Long before humans began to sow harvested seeds, they practiced various forms of landscape management, for instance, cultivating wild cereals and pulses by tilling the soil. In the way that we dealt earlier with proto-languages, we will now deal with predomestication cultivation. Unlike fully developed agriculture, predomestication cultivation, in the sense of the manipulation of wild plants and animals, did not in itself constitute a complete subsistence strategy but only one component of such a strategy. It evidently existed for a very long time in human history but played only a more or less marginal role for food production, in the same way that early communication systems must have initially played a rather marginal role in human cooperation. And this role was certainly not motivated by the later outcomes of domestication but constituted an activity with its own rationale and dynamics. Predomestication cultivation offers an example of the principle mentioned earlier that the horizon of applications of given means is always larger than the intentions for which they had been originally employed. This may even apply literally to some of the instruments employed in early farming, which had earlier been used for other purposes.

At least in the Fertile Crescent there were several reasons why predomestication cultivation did not remain marginal, in particular, the contingent ecological conditions that encouraged sedentariness. Sedentariness favored the extension of cultivation practices bound to local environments. Given the investment of labor in cultivation practices, such local predomestication cultivation practices in turn stabilized sedentariness, thus creating what Dorian Fuller has aptly called the "labor traps" along the protracted trajectories leading to domestication (Fuller et al. 2010, 2011). This mutual reinforcement of sedentariness and cultivation is similar to the stabilization effect of the development of a prelinguistic communicative system that was pointed out earlier. It constitutes a kind of resonance effect between external conditions and the internal structure of the evolving system. In any case, there was initially no guarantee that predomestication cultivation would lead necessarily to domestication proper. Only at some points along some trajectories may "tipping points" (Fuller) have been reached that then drove the further development in a particular direction, whereas other trajectories may have been aborted or remained in intermediate stages. Contingent external circumstances had thus been transformed into conditions for the internal stability and further development of a society.

We can also see an analogue to the process of decontextualization in the internalization of external conditions mentioned earlier for language evolution. Eventually, domesticated crops were no longer bound to the local contexts in which their ancestors were originally found but spread into other areas and ultimately across the world. Such globalization effects—also important to the evolutionary history of languages—may have helped to emancipate the incipient domestication processes from the variety of local contexts in which they took place. Since cultivation was part of a network activity taking place in an extended geographical area (and not just in a small core region as traditionally assumed), migration and exchange among different sedentary communities eventually contributed to a diversification and enrichment of cultivars at any specific location. The resulting recontextualization of cultivation may also have helped to separate wild from cultivated populations, thus contributing to a process by which human-defined plant or animal populations were ultimately transformed into biologically defined populations. We also briefly mention here another element of the co-evolutionary and niche construction dynamics of the Neolithic Revolution, namely the co-evolution of disease and society in the context of emerging trade networks and agricultural practices. The emerging constructed niche of agricultural and sedentary societies reaching new levels of population density facilitated the evolution of a number of infectious diseases, which in turn had a huge effect on the regulative structures of these societies (Diamond 1998; McNeill 1976).

9.7 The Evolution of Writing Systems

Our third example, based on the work of Peter Damerow, deals with what has often been called the urban revolution that is associated with the emergence of writing in southern Mesopotamia in the second half of the fourth millennium BC (Damerow 2012, Renn 2015b). Urbanization is of course based on the Neolithization process we have just described. In this period the modest accounting techniques that had been developed earlier in the context of the rural economy of Babylonia were extensively exploited in the administrations of emerging city-states. These modest accounting techniques correspond to the predomestication cultivation practices in our previous example and to precursors of linguistic communication in our first example. Among the traditional accounting techniques were small clay tokens of different shapes, which served as symbolic representations of objects and were used for representing and controlling their quantities, but also seals representing certain administrative acts. The exploration of these given means, which served as external representations of administrative knowledge, in the context of an expanding economy eventually led to a transformation of the traditional symbolic culture. The potential of existing tools for symbolic representation was exploited to its limits, with the effect of stabilizing the economies of the emerging city-states. The exploration of the potential of these tokens led, for instance, to a proliferation of the number and shapes of the clay counters, which had originally been used only in small quantities in the context of rural communities.

A critical turning point was when these two elements of the traditional accounting techniques—the counters used for keeping track of the quantities of the administered objects and the seal impressions documenting administrative acts—were represented within a single medium: the clay tablets serving as the earliest medium of writing. Two initially separate accounting techniques thus became integrated in a new form of external representation whose enormous potential was explored in the sequel. The so-called numero-ideographic tablets, representing an early stage of proto-writing, became the starting point for the exploration of new forms of information storage and processing in the archaic period of Babylonian society. The tablets could hold more information than the earlier administrative techniques and that information could be more flexibly and efficiently structured. As in the evolution of language, we thus see that the exploration of a system of representation could act as the population-level analogue of Vygotsky's zone of proximal development.

The next step in the development of writing was again shaped by the fundamental property of external representations in that the range of their possible applications was larger than the specific goals for which they had initially been introduced. The potential of early proto-writing to represent mental constructions in fact reached far beyond the limited field of application within Babylonian administration for which this technique had originally been introduced. In its most developed form, achieved after about a 1000 years of historical development, it also included the possibility to represent spoken language. Writing in the modern sense thus emerged from a representational system that was originally developed without this goal in mind, just as domestication resulted from cultivation practices originally pursued without this aim. Such novel possibilities typically occur only as a side effect of mainstream applications. And it is also characteristic that the foundational role of these marginal applications as being constitutive of a new stage is only realized once a new perspective is introduced, often triggered by a new external context.

One such context was education. Indeed, the growing complexity of the proto-writing system required institutional support for its transmission from generation to generation. But schooling implies a separation of the cognitive means of administration from their immediate context of application and thus opens up a perspective in which the potential of these cognitive means could be explored independently from the constraints of their application to solve concrete administrative problems. The role of education provides a good example for the emancipation of a system of knowledge from its embedding within concrete contexts of application. But there were also other factors that may have similarly acted toward a recontextualization of the existing system of proto-writing, such as the spread of the system across cultural boundaries. Thus a more reflective perspective on this system was introduced, favoring the discovery that it was possible to repurpose proto-writing to represent language.

9.8 The Dynamics of Cultural Evolution

Against the background of these three very different examples we can now summarize our model of cultural evolution. What we have been describing are networks of human actions that include a given material and social culture. This cultural niche results from prior actions and constitutes, together with other environmental conditions and the internal organization of the actors, the regulative structure governing the network. The evolution of a networked population consists in its dealing with outside challenges to its preservation and that of its systemic structures. In the process, such external challenges may be internalized, that is, transformed into elements of the internal network structure.

The cognitive structures of individual actors are generated by reflecting on environmentally embedded actions. They represent the cognitive dimension of an action potential that we designate as knowledge. Knowledge itself may be externally represented and is thus shareable within a knowledge economy. While innovations may be triggered by external challenges, they become possible first because the horizon of possibilities associated with given means and external representations is always larger than the goals pursued by any given set of actors, and second because the context or the results of actions may become a source of new means of action and new external representations which then enable new forms of action coordination.

The evolution of a networked population may lead to the establishment of new patterns of interaction and new forms of internal, cognitive organization of the actors. On the individual level, new regulative structures arise because the reflection on actions with external representations may generate knowledge of a higher order of abstraction than the knowledge to which the external representation originally referred. The results of such reflections may then again be externally represented, thus generating context and path-dependent chains of abstractions of increasing order. On the population level, new regulative structures may emerge from the introduction of new means or from the exploration of given means, typically triggering both new forms of social organization and new forms of knowledge. Such new patterns of interaction are typically layered in the sense that the introduction of a new pattern does not lead to the complete eclipse of earlier patterns, which are rather integrated into the subsequent layers.

9.9 Outlook

Given the subject of this volume we have chosen our examples mainly from the domain of cultural evolution. But the conceptual framework we suggest is more general and provides a new perspective on all evolutionary processes—biological and cultural. To briefly illustrate how an integrated perspective on regulatory networks and niche construction applies more generally, we conclude with one particular example from biology: the evolution of eusocial insects (Laubichler and Renn

2015; Hölldobler and Wilson 2009; Page 2013). The evolution of these complex societies has long puzzled evolutionary theorists, mainly because of their remarkable reproductive division of labor, with a majority of the individuals not reproducing at all. But how can natural selection, favoring reproductive success, lead to giving up one's reproductive potential? In the last decades theorists have focused on models, either kin or group selection, to address this question. But this is not the crucial question. The more fundamental question is: how do we get any division of labor and differentiation into different caste types? And how does this work, given that all individuals in a colony have more or less the same genome? And how can such systems be the product of evolution?

The problem of the evolution of such a superorganism is the same as the problem of the evolution of any multicellular organism with different cell types, namely the problem of developmental regulation and niche construction (Laubichler and Renn 2015). Organisms (super or normal) develop from simple beginnings (a fertilized egg or a mated queen) and gradually differentiate into complex systems (a complex multicellular organism or a fully differentiated colony). At each step this process is controlled by a regulatory system, anchored in the genome, but extending outward and closely connected to co-constructed niches. This regulatory apparatus—more than 90% of the genome—processes input, both internal and external, and assures a predictable developmental trajectory as well as the maintenance of system states.

For example, ant colonies have a fixed percentage of soldiers. Removing those—changing the niche for the brood caregivers—triggers a signal cascade that begins at the level of social interactions (reduced encounters with soldiers), then goes to the brain that processes the social signals and changes the behavior accordingly. The freshly laid larvae are now fed differently, which then triggers physiological processes and cellular signal cascades that ultimately change gene expression profiles so that soldiers hatch instead of other caste types. And this happens until the normal ratio of soldiers is restored. Not all the information of this regulatory system thus resides inside the genome; rather the various niches provide crucial regulatory effects, which are themselves the product of co-evolutionary dynamics. To highlight the inclusion of this external part into an extended regulatory system we refer to this phenomenon, which may also have played a role at the origin of life, as "endaptation" in contrast to the familiar concept of exaptation which designates the repurposing of existing characters.

These remarkable phenomena of differentiation are a consequence of the regulatory developmental potential of complex inheritance systems, including genomes and their external parts, which, as we now know, greatly influence the course of evolution by determining what kind of variation can be introduced into the system. This developmental evolution perspective accounts for the path-dependent and thus historical nature of evolution by analyzing the properties of complex inheritance systems and their interactions with a nested layer of co-constructed niches, from genomes to cultures. It focuses on explaining the origin of variation as a consequence of the properties of these complex systems (Laubichler and Renn 2015).

Evolutionary theory is thus no longer based on "change in allele frequencies" but follows the same logic of extended evolution that we have summarized on the basis of our cases of cultural evolution. Our perspective, based on networks and contexts, hopefully overcomes Mach's worry about the relation between biological and cultural evolution:

> Here I wish simply to consider the growth of natural *knowledge* in the light of the theory of evolution. For knowledge, too, is a product of organic nature. And although ideas, as such, do not comport themselves in all respects like independent organic individuals, and although violent comparisons should be avoided, still if Darwin reasoned rightly, the general imprint of evolution and transformation must be noticeable in ideas also. (Mach 1986, 218)

Acknowledgements This paper is based on a talk given by one of us (J.R.) at the international conference "Integrated History and Philosophy of Science – &HPS5," held at the University of Vienna from June 26–29, 2014. We are grateful to Friedrich Stadler for giving us the opportunity to present our approach on this occasion. The paper is based on earlier publications of the authors on related themes listed in the references. We would like to thank Sascha Freyberg, Matthias Schemmel and Matteo Valleriani for helpful discussions from which many of the ideas presented here have emerged. We are especially grateful to Lindy Divarci who carefully edited the text.

References

Bowles, Samuel, and Herbert Gintis. 2011. *A Cooperative Species: Human Reciprocity and Its Evolution*. Princeton: Princeton University Press.

Damerow, Peter. 1996. *Abstraction and Representation: Essays on the Cultural Evolution of Thinking*, Boston Studies in the Philosophy of Science. Vol. 175. Dordrecht: Kluwer Academic Publishers.

———. 2000. How Can Discontinuities in Evolution Be Conceptualized? *Cultural Psychology* 6 (2): 155–160.

———. 2012. The Origins of Writing and Arithmetic. In *The Globalization of Knowledge in History*, Studies 1: Max Planck Research Library in the History and Development of Knowledge, ed. Jürgen Renn, 153–173. Berlin: Edition Open Access.

Dediu, Dan, and Stephen C. Levinson. 2013. On the Antiquity of Language: The Reinterpretation of Neandertal Linguistic Capacities and Its Consequences. *Frontiers in Psychology* 4: 397.

Diamond, Jared M. 1998. *Guns, Germs, and Steel: The Fates of Human Societies*. New York: Norton.

Fuller, Dorian Q., Robin G. Allaby, and Chris Stevens. 2010. Domestication as Innovation: The Entanglement of Techniques, Technology and Change in the Domestication of Cereal Crops. *World Archaeology* 42 (1): 13–28.

Fuller, Dorian Q., George Wilcox, and Robin G. Allaby. 2011. Cultivation and Domestication Had Multiple Origins: Arguments against the Core Area Hypothesis for the Origins of Agriculture in the Near East. *World Archaeology* 43 (4): 628–652.

Hölldobler, Bert, and Edward O. Wilson. 2009. *The Leafcutter Ants: Civilization by Instinct*. New York: W.W. Norton.

Jeffares, Ben. 2012. Thinking Tools: Acquired Skills, Cultural Niche Construction, and Thinking with Things. *The Behavioral and Brain Sciences* 35 (4): 228–229.

Laland, Kevin N., et al. 2000. Niche Construction, Biological Evolution, and Cultural Change. *The Behavioral and Brain Sciences* 23 (1): 131–146; see also the discussion on pp. 146–175.

Laland, K. N., Odling-Smee, J. and Gilbert, S. F. (2008), EvoDevo and niche construction: building bridges. J. Exp. Zool., 310B: 549–566. doi:10.1002/jez.b.21232.

Laubichler, Manfred, and Jürgen Renn. 2015. Extended evolution: A conceptual framework for integrating regulatory networks and niche construction. *JEZ Part B: Molecular and Developmental Evolution* 324 (7): 565–577.
Levinson, Stephen C. and Judith Holler. 2014. The Origin of Human Multimodal Communication. *Philosophical Transactions of the Royal Society B, 369: 20130302.*
Lock, Andrew J. 2000. Phylogenetic Time and Symbol Creation: Where Do Zopeds Come from? *Culture & Psychology* 6 (2): 105–129.
Mace, Ruth, and Clare J. Holden. 2005. A Phylogenetic Approach to Cultural Evolution. *Trends in Ecology & Evolution* 20 (3): 116–121.
Mach, Ernst. 1986. *Popular Scientific Lectures*. La Salle: Open Court.
———. 2011. Erkenntnis und Irrtum. In *Skizzen zur Psychologie der Forschung. Ernst Mach Studienausgabe*. Berlin: Xenomoi Verlag.
Maynard Smith, John, and Eörs Szathmáry. 1995. *The Major Transitions in Evolution*. Oxford: Oxford University Press.
McNeill, William. 1976. *Plagues and Peoples*. New York: Anchor Books Doubleday.
Mesoudi, Alex, et al. 2004. Perspective: Is Human Cultural Evolution Darwinian? Evidence Reviewed from the Perspective of the Origin of Species. *Evolution* 58 (1): 1–11.
———. 2011. *Cultural Evolution: How Darwinian Theory Can Explain Human Culture and Synthesize the Social Sciences*. London: University of Chicago Press.
Odling-Smee, John. 1995. Niche Construction, Genetic Evolution and Cultural Change. *Behavioural Processes* 35 (1–3): 195–205.
Page, Robert E. 2013. *The Spirit of the Hive: The Mechanisms of Social Evolution*. Cambridge, MA: Harvard University Press.
Read, Dwight W. 2014. The Substance of Cultural Evolution: Culturally Framed Systems of Social Organization. *The Behavioral and Brain Sciences* 37 (3): 270–271.
Renn, Jürgen, ed. 2012. *The Globalization of Knowledge in History*, Studies 1: Max Planck Research Library in the History and Development of Knowledge. Berlin: Edition Open Access.
———. 2014. The Globalization of Knowledge in History and its Normative Challenges. *Rechtsgeschichte/Legal History* 22: 52–60.
———. 2015a. From the History of Science to the History of Knowledge – and Back. *Centaurus* 57: 37–53.
———. 2015b. Learning from Kushim about the Origin of Writing and Farming. In *Grain Vapor Ray. Textures of the Anthropocene*, ed. K. Klingan, A. Sepahvand, C. Rosol, and B.M. Scherer, 241–259. Cambridge, MA: MIT Press.
Richerson, Peter J., and Robert Boyd. 2005. *Not by Genes Alone: How Culture Transformed Human Evolution*. Chicago: University of Chicago Press.
Richerson, Peter J., and Morten H. Christiansen. 2013. *Cultural Evolution: Society, Technology, Language, and Religion*, Strungmann Forum Reports. Cambridge, MA: MIT Press.
Smith, Leslie, Julie Dockrell, and Peter Tomlinson. 1997. *Piaget, Vygotsky and Beyond: Future Issues for Developmental Psychology and Education*. New York: Routledge.
Tomasello, Michael. 2014. *A Natural History of Human Thinking*. Cambridge, MA: Harvard University Press.
Vygotsky, Lev S. 1978. Mind in Society. In *The Development of Higher Psychological Processes*. Cambridge, MA: Harvard University Press.
Wimsatt, W.C. 2013. Articulating Babel: An Approach to Cultural Evolution. *Studies in History and Philosophy of Biological and Biomedical Sciences* 44 (4): 563–571.

Part II
General Part

Chapter 10
Carnap's *Weltanschauung* and the *Jugendbewegung*: The Story of an Omitted Chapter

Adam Tamas Tuboly

10.1 Introduction

Richard Creath (2007, 332) claimed earlier that "Quine did arrive in Vienna in 1932, but intellectually, at least, he never left. [...] Vienna remained the city of Quine's dreams; it was the home of his concerns, the source of his arguments, and the lode-star of his aspirations." If Vienna was the city of Quine's dreams, then it was indeed the city of Rudolf Carnap.

According to the usual story, after Carnap arrived in Vienna in 1926 (first in 1925 to present the *Aufbau* as his *Habilitationsschrift*), he found himself in a stimulating and cooperative atmosphere. For Carnap, originally a physicist, who tried to explicate the connections between physics, mathematics, logic, and philosophy while searching for a general and unified scientific framework, Vienna offered the required help both to finish his ongoing projects and to conceptualize the further scientific-philosophical works.

Even the Vienna Circle welcomed Carnap as the long-awaited system-builder who could synthesize their various efforts and philosophical insights into a general framework which would connect all the dots. As Philipp Frank (1949, 33) put it: "[In the *Aufbau*] the integration of Mach and Poincaré was actually performed in a coherent system of conspicuous logical simplicity. Our Viennese group saw in Carnap's work the synthesis that we had advocated for many years." But even if we do not take at face value the retrospective – and as Thomas Uebel (2003) said – highly "programmatic" historiography of Frank, already in 1929 the authors of the Circle's manifesto (Carnap, Neurath, and partly Hahn, Feigl and Waismann) claimed that (in the context of their method):

A.T. Tuboly (✉)
Hungarian Academy of Sciences, Budapest, Hungary
e-mail: Tuboly.Adam@btk.mta.hu

F. Stadler (ed.), *Integrated History and Philosophy of Science*, Vienna Circle Institute Yearbook 20, DOI 10.1007/978-3-319-53258-5_10

[i]f such an analysis were carried through for all concepts, they would thus be ordered into a reductive system, a '*constitutive system*'. Investigations towards such a constitutive system, the '*constitutive theory*', thus form the framework within which logical analysis is applied by the *scientific world-conception*. (Carnap et al. 1929/1973, 309. Italics added.)

Thus before Neurath's critique and Carnap's physicalist turn, a part of the Circle maintained that Carnap's general ideas about concept-building (in the *Aufbau*) provided the required framework to spell out their (scientific) conception of the world.

Vienna seemed to be, however, the city of Carnap's dream from a broader cultural perspective too. The manifesto's authors (among them Carnap) said that

[i]n the second half of the nineteenth century, liberalism was long the dominant political current. Its world of ideas stems from the enlightenment, from empiricism, utilitarianism and the free trade movement of England. In Vienna's liberal movement, scholars of world renown occupied leading positions. Here an anti-metaphysical spirit was cultivated [...]. (Carnap et al. 1929/1973, 301.)

Besides the diverse scientific landscape, Vienna showed a colorful picture of political, social and cultural ideas. Even Carnap seems to confirm that Vienna was an ideal place for him after he visited the Circle for the very first time in 1925: "Besides[the philosophical atmosphere] Vienna is attractive too: a lot of cultures, a lot of international lives."[1]

So far so good one could say, concerning, at least, the usual story. As a part of Carnap's *Nachlass*, however, in his original and unpublished intellectual autobiography written in the late 1950s for the Schilpp volume, Carnap draws our attention to a quite different narrative of his "Wiener Projekt"[2]:

After the war [...] the same spirit was still alive [vivid] in the life of my newly founded family and in the relationships with friends. When I went to Vienna, however, the situation changed. I still preserved the same spirit in my personal attitude, but I missed it painfully in the social life with others. None of the members of the Vienna Circle had taken part in the Youth Movement, and I did not feel myself strong and productive enough to transform singlehandedly the group of friends into a living community, sharing the style of life which I wanted. Although I was able to play a leading role in the philosophical work of the group, I was unable to fulfill the task of a missionary or a prophet. Thus I often felt as perhaps a man might feel who has lived in a religious[ly] inspired community and then suddenly finds himself isolated in the Diaspora and not strong enough to convert the heathen. The same feeling I had in a still greater measure later in America, where the power of traditional social conventions is much stronger than it was in Vienna and where also the number of those who have at least sensed some dissatisfaction with the traditional forms of life is smaller than anywhere on the European continent. (Carnap 1957, [UCLA] Box 2, CM3, folder M-A5, pp. B35–B36.)

This passage is purported to show that even though Vienna could have been the city of Carnap's dreams from a theoretical (philosophical and scientific) point of view,[3]

[1] Carnap to Reichenbach, March 10, 1925. ASP RC 102-64-11. All translations are mine.

[2] The term is from Carnap's letter to his father-in-law. It is dated just after Carnap went back to Wiesneck after his visit to Vienna, November 2, 1925. ASP RC 102-23-01.

[3] As many scholars argue, even from a philosophical point of view, Vienna could not cover the whole interest of Carnap due to the anti-Kantian tendencies of Austrian philosophy. In 1933

from a broader cultural (social and political) perspective the Viennese people just missed something important and fundamental: none of them have taken part in the so-called German Youth Movement (GYM), the *Jugendbewegung*.

The role and lasting effect of GYM on Carnap's thought and philosophy were emphasized recently, for example, by Gottfried Gabriel (2004), André Carus (2007a, b), Christian Damböck (2012) and Jacques Bouveresse (2012). The aim of this paper is to make some further comments on Carnap's relation to the GYM, particularly on the question of why was it omitted from his published intellectual autobiography?

I will proceed as follows. In Sect. 10.2 the *Jugendbewegung* is going to be discussed, particularly its effect on Carnap's *Weltanschauung*. Then in Sect. 10.3 I will present some reasons which led finally to the decision to cut from the autobiography those passages which concerned the *Jugendbewegung*.

One could naturally raise the question whether such a micro-story about an omitted chapter is important at all. Since I claim that (at least partly) the GYM's effect could be detected both in the principle of tolerance and in Carnap's general metaphilosophy, it indeed seems to be relevant to deal with the omitted passage and its context. On the other hand, since the GYM was not present in Carnap's philosophy as a propositionally formulated piece of knowledge (he never refers to it as such), approaching the problem from the idea of worldviews gives us a proper framework. According to Wilhelm Dilthey (1968, 78), "the deepest root of *Weltanschauung* is in life itself" thus we shall reveal those socio-cultural moments which made possible and framed Carnap's views. GYM was just such a moment. Furthermore, as Karl Mannheim (1921–22/1959, 45) formulated it, the analyzes of worldviews and cultural objects "embraces not merely cultural products endowed with traditional prestige, such as Art or Religion, but also manifestations of everyday life which usually pass unnoticed [...]", like participation in a movement.

10.2 The *Jugendbewegung* and Its Effect on Carnap's *Weltanschauung*

In an interview (Haller and Rutte 1977, 27–28), Heinrich Neider, a former member of the Circle, remembered Carnap as follows:

> [Carnap] was then [around World War I] an independent social democrat [...], Carnap was never a communist. But he was nevertheless a radical socialist, even if it was not something you would have guessed when you saw him. He was a man unable of any outburst of affect,

Carnap wrote a short intellectual autobiography to Marcel Boll, in which he said: "It is characteristic of the recent German philosophical situation that as a German of the Reich [Reichsdeutscher] I found my field of activity [Tätigkeitsfeld] not in Germany but in Vienna and Prague [...]." ASP RC 091-20-09. One could interpret this passage as Carnap tries to give voice to his dissatisfaction that he had to leave Germany (even though he has found himself in a fruitful and cooperative atmosphere among logical empiricist outside of Germany).

> from whom I have never heard an impolite or despising word and to whom any kind of fanaticism was alien. I considered him which such a reaction of incredulity, when he said: 'I, who was an independent at that time', and I said: 'I would absolutely not believe that of you' and he answered to that by the following reflection: 'There are many things you would not believe about me, I have also been there at the *Hohe Meißner* festival'.[4]

The "Hohe Meißner" is a mountain in Hesse, Germany, where Germans planned to celebrate the Battle of Leipzig against Napoleon. 1913 was the centenary and they organized a huge national-military-patriotic festival. On the 11th and 12th of October, 1913 the members of the GYM planned a huge counter-festival, with 4000 participants from all over the country; different groups of the Movement were gathering at the top of the mountain. One enthusiastic member and actually organizer of the counter-festival was Carnap.[5]

The GYM, whose first group was called the *Wandervogel* [birds of passage], began at the end of the nineteenth century in Berlin[6]: it was a "large-scale rebellion of well-off adolescents against the perceived conformism of their parents and teachers to the rigid norms of the society into which students were being socialized" (Carus 2007a, 50). The main roots of the GYM could be found in German Romanticism but members of the GYM tried to revive some customs and habits also from medieval times: they arranged extensive and long ramblings in the countryside, where they eat what they find and could make from the elements of nature.[7] They tried to get closer to the peasantry and master their lifestyle with all its naiveté and purity.

The latter characteristics were of the utmost importance for the participants. Members of the GYM abstained from the "bourgeois" vices and drugs, such as coffee, tobacco, alcohol. As Quine (1971, xxiv) recalled later: "Carnap's habits were already austere: no science after dinner, on pain of a sleepless night. No alcohol ever. No coffee." So instead of the usual contemporary lifestyle or traditions from the city, these young people created their *own habits* and *culture*: they sang while the walked, slept under the open sky, danced and read poems.

---

[4] Translated by Jacques Bouveresse (2012, 56).

[5] Another participant was Hans Reichenbach with a delegation of the *Freistudenten* [Free Students] from Berlin. Earlier Reichenbach was also a member of the *Wandervogel* movement and later took an active part in the *Freie Studentenschaft*. See the memoir of Carl Landauer (1978). Reichenbach's experiences in the GYM had a similar effect on his thought as on Carnap's. Kamlah (2013) provides a detailed analysis of Reichenbach's volitional conception of ethics and decisions regarding both philosophy and science. I am indebted to Thomas Uebel for calling my attention to the case of Reichenbach.

[6] About the GYM see Laqueur (1962), Aufmuth (1979); Bias-Engels (1988) and Werner (2003).

[7] As Laquer (1962, 15–16) said "[…] the early *Wandervogel* put itself into deliberate opposition to a society whose interest in nature was by and large limited to yearly visits to mountain or seaside resorts, with all their modern comforts. There was more to it, too. It was, or at any rate became, a somewhat inchoate revolt against authority."

Carnap was a member of the GYM's local group in Jena, called Serakreis [Sera Circle]; it was organized by the famous publisher Eugen Diederichs.[8] Actually, Carnap was present at the Hohe Meißner as one of the representatives and for some time leader of the Serakreis. He remembered the gatherings of the circle, especially its Festival of the Midsummer night as follows:

> Influenced by Skandinavian customs, there were songs, dances, and plays. Diederichs read the Hymn to the Sun by St. Francis of Assisi, after sundown the big fire was lighted, encircled by the large chain of singing boys and girls, and when the fire had burned down there came the jumping of the couples through the flames. Finally, when the large crows of guests had left, our own Circle remained at rest around the glowing embers, listening to a song or talking softly, until we fell asleep in the quiet night under the starry sky. (Carnap, 1957, [UCLA], Box 2, CM3, MA-5, p. B30.)

The aim of the movements was "to find a way of life which was genuine, sincere, and honest, in contrast to the fakes and frauds of traditional bourgeois life; a life, guided by the own conscience and the own standards of responsibility and not by the obsolete norms of tradition."[9] Though Carnap complained a lot about his memory[10] and the autobiography is indeed inaccurate and sloppy at some points, his recollection about the GYM agrees with the original documentation of the movement: "Free German Youth desires, of its own determination and under its own responsibility, to shape its life with inner authenticity [Wahrhaftigkeit]. It stands united for this inner freedom under all circumstances" (Messer 1924, 19–20).

It would be hard to overestimate the influence of the GYM on Carnap's thought. In the unedited and unpublished intellectual autobiography, he even devoted a section to these ideas, entitled as "Weltanschauung: Religion, enlightenment, youth movement" (B18). What he learned and acquired there is not a set of theoretical statements and doctrines, but a way and attitude towards life [Lebensgefühl], a form of life [Lebensform] and a certain worldview [Weltanschauung]. One could say with Dilthey (1968, 86) that worldviews are not the "products of thought." A worldview is, after all, such an a-logical, non-conceptual and non-structured totality of feelings and experiences which underline *all* the products of the human mind [Geist]. From such a viewpoint, "theoretical philosophy is neither the creator nor the principal vehicle of the *Weltanschauung* of an epoch; in reality, it is merely only one of the channels through which a global factor – to be conceived as transcending the various cultural fields, its emanations – manifests itself" (Mannheim 1921–22/1959, 38). Philosophical contents, considered as cultural products and philosophical styles, are just expressions and documentations of worldviews.

Since worldviews are pre-propositional, they are evidently having a non-theoretical character; but they are not irrational if we mean by the concept something *meaningless*. Worldviews are rather a-theoretical (and/or a-rational) complexes of feelings and experiences, hence rational justification is not required in their case:

---

[8] Didereichs was an important figure later too: as a publisher he published the books of Franz Roh, Wilhelm Flitner and Walter Fränzel, who were close collaborators of Carnap in the early 1920s.

[9] Carnap, 1957, [UCLA], Box 2, CM3, MA-5, p. B31–B32.

[10] See for example his letter to Brian McGuinnes, November 27, 1969. ASP RC 027-33-14.

worldviews do not violate the rules and norms of rationality since they serve as the hidden, but the continuous base of rationality and theoretical argumentation.[11]

Though Carnap is evidently not referring to what has been said earlier, one could still interpret his words as claiming that the suitable cultural medium and social experiences could influence philosophy itself in a fruitful manner, which is, as Mannheim (1921–22/1959, 38) said, "merely one of [the] manifestations [of world-views] and not the only one":

> For those whose work is on a purely theoretical nature, there is the danger of a too narrow concentration on the intellectual side of life, so that the properly human side may be neglected. I think it was very fortunate for my personal development during these decisive years that I could participate both in Freiburg and in Jena in the common life of such fine and inspired groups of the Youth Movement. (Carnap, 1957, [UCLA], Box 2, CM3, MA-5, p. B32.)

Though Carnap participated in the GYM only between 1910 and 1914, he actively maintained his relation to his fellows during and after the First World War.[12] He continuously corresponded with the members, read their pamphlets and articles which were published in their journals. His friendships made in the movement turned out to be lasting for decades and in some cases, they were life-long relations. Carnap got to know the German sociologist Hans Freyer in the GYM, and Freyer's ideas about the *Geisteswissenschaften* became very influential on the *Aufbau* and Freyer played an important role in transmitting the philosophy of Wilhelm Dilthey in the formative years of Carnap.[13] After the First World War, Carnap organized a workshop in Buchenbach about the "system of sciences": the participants were his closest friends from the Serakreis, namely Freyer, the pedagogic Wilhelm Flitner, and the art historian and photographer Franz Roh. The discussion group of Carnap, Freyer, Roh and Flitner in the summer of 1920 had a well-documented effect on the *Aufbau* and on the early thoughts of Carnap.[14]

All these friends shared the same experiences in the GYM and the movement's impact remained quite detectable and fundamentally important for Carnap:

[11] According to Tamás Demeter (2012, 49), worldviews could be approached as a form of Kantian conditions of possibility, especially like the forms of intuition. Neither of them have a conceptual character, they do not mean knowledge, they do not possess a propositional structure but they still make possible knowledge in a broader sense: "[w]e could say in the Kantian idiom, *Weltanschauung* is empirically real but transcendentally ideal: works of cultural production are impossible independently of a worldview, but a worldview cannot be known independently of the works of cultural production."

[12] Actually Carnap received two letters from Martha Hörmann, a former member of the Serakreis, in 1964. She told Carnap about the 1963 meeting at the Hohe Meißner, and how it revived her feelings and memories from the formative years in Jena. See ASP AC 027-29-26 and ASP RC 027-29-27.

[13] About Dilthey's indirect influence on Carnap's thought see Gabriel (2004); Damböck (2012); Tuboly (forthcoming).

[14] Freyer and Flitner were also members of the GYM and while Carnap's friendship with Freyer broke in the early 1930s when Freyer moved to the political right, Flitner, Roh and Carnap were life-long friends. About the Buchenbach-conference see Dahms (2016), about Flitner's recollection see Flitner (1986).

> [...] the spirit that lived in this movement, which was like a religion without dogmas, remained a precious inheritance for everyone who had the good luck to take an active part in it. What remained was more than a mere reminiscence of an enjoyable time; it was rather an indestructible living strength which forever would influence one's reactions to all practical problems of life. (Carnap, 1957, [UCLA], Box 2, CM3, MA-5, p. B34–B35.)

What Carnap learned is a certain *attitude*: we should not accept blindly any doctrine, knowledge and the heritage of our ancestors and other authorities just as it stand. We have the right and ability to revise everything, to reshape and rebuild (Aufbau) our material, cultural and social environment and to question every convention and arrange our cultural world as we wish. We have a total freedom [völlige freiheit][15] in these questions. Carnap formulated these ideas in his published autobiography under the label of "scientific humanism":

> [...] man has no supernatural protectors or enemies and that therefore whatever can be done to improve life is the task of man himself. [...] we had the conviction that mankind is able to change the conditions of life in such a way that many of the sufferings of today may be avoided and that the external and the internal situation of life for the individual, the community, and finally for humanity will be essentially improved. (Carnap 1963, 83)

In his unpublished autobiography, actually, he told a story about a conversation with a peasant in a remote village of the Black Forest after the First World War which documents the above-mentioned trends:

> We looked at an airplane at great distance, high in the sky, and he said: 'They say that sometimes people fly in such machines. But that is not possible.' I told him that I had flown a few times in an airplane. He looked at me somewhat suspiciously, shook his head, and said: 'Now look here: I am much older than you; I know very well what can be done and what cannot. Now you believe me, this thing is just not possible.' (Carnap 1957, [UCLA], Box 2, CM3, MA-4, pp. N17–N18.)

This example shows quite well that attitude against which Carnap and his youth friends stood up.

I would like to end this section with the mentioning of the examples where one can evidently find the effect of the GYM on Carnap's thought. First there is the notorious *principle of tolerance* (Carnap 1934/1937, §17), which says, after all, that one is totally free to choose between logical systems and (philosophical/scientific) languages as he wishes (though the principle was extendable for methods also). Engineer your schemes and conceptions as you wish, decide which one fits your space of (practical and theoretical) reasons the best and leave behind the authoritative *a priori*/armchair-style philosophical reasoning.

The second point (which is actually connected to the first) is that our freedom is extended also to the practical realm through the dialectical conception of explication (Carus 2007a): since the possible consequences of the various possible acts affect our practices and these consequences are codified in different language forms, our actions and practical decisions are not fixed but relative in a sense to a particular language form. This conception was formulated compactly by Richard Jeffrey

[15] Carnap used these words when he introduced his principle of tolerance in the discussions of the Circle in 1933. See ASP RC 110-07-22.

(1994, 847) who was a close collaborator of Carnap on the theories of probability and inductive logic in the last few decades of his life:

> Philosophically, Carnap was a social democrat; his ideals were those of the enlightenment. His persistent, central idea was: »It's high time we took charge of our own mental lives«, time to engineer our own conceptual scheme (language, theories) as best we can to serve our own purposes; [...] time to accept the fact that there's nobody out there but us, to choose our purposes and concepts to serve those purposes, if indeed we are to choose those things and not simply suffer them. [...] For Carnap, deliberate choice of the syntax and semantics of our language was more than a possibility it was a duty we owe ourselves as a corollary of freedom.

If the GYM had such a detectable and important influence on Carnap's intellectual development as claimed here, then one could rightly ask that why did he cut it from his intellectual autobiography? I will try to indicate some possible reasons in the next section.

10.3 The "Intimate" Parts of the Biography

In the recent literature on Carnap, it is frequently emphasized that his original intellectual autobiography written for the Schilpp volume in the second half of the 1950s was much longer and detailed than the published one in 1963.[16] Carnap cut his autobiography but there were certain shortenings also in his replies and systematic presentations of his philosophical ideas. So far Carnap-scholars did not focus on the reasons for this move besides that it was due to the unexpected length of the volume which, at some point, was considered to be published in two volumes just because of that.

Carnap's unpublished correspondence, however, is promising in this context and I will concentrate on three points. These points (or reasons) have in a general (and neutral) sense suitable documentary-value: they express aptly Carnap's worldview, the idea that the fallible and contingent factors of "everyday life" could bear on theoretical issues, and the trends of the social and political epoch of his time.

In a broader sense, we could connect these points and Mannheim's (1921–22/1959, 44) inquiries about the "strata of meaning" of the cultural and social products. Mannheim claims that we can differentiate three levels: (a) the objective meaning, (b) the intentional-expressive meaning, and the (c) documentary or evidential meaning. In the first case, to understand a given act, we shall abstract from the participant subjects, from their intentions and psychological states, and it is

[16] It seems, in fact, that the Schilpp volume, *The Philosophy of Rudolf Carnap*, did not appear in 1963. Carnap wrote to Robert Mathers on 20 November, 1963 that "I hear that the Schilpp volume is to appear by Dec. 31; but there have been so many delays that this date cannot be counted on" (ASP RC 088-62-09). Still on 5 April, 1964, Carnap told to Albert Blumberg that "Unfortunately I am not able to send you a copy of my autobiography. In the mistaken trust that the Schilpp-volume was going to appear in 1963 I seem to have given away all my copies" (ASP RC 088-06-04).

enough to know the "objective social configuration" (ibid. p. 45). Only in a given social configuration will a bit of metal function as alms. In the second case, the subject will be relevant and the meaning of the act "cannot be divorced from the subject and his actual stream of experience, but acquires its fully individualized content only with reference to this 'intimate' universe" (ibid. p. 46). What matters is what the subject intends to *express* with the given act. Finally, in the third case, besides the social configurations, and the intended expressive elements, the important thing is "what is documented about [the subject], albeit unintentionally, by that act of his" (ibid. p. 47).

Given that Carnap's act, namely the cutting of the passages about the GYM, could be considered as a cultural and social product of the human mind [Geist], we could use Mannheim's approach. Actually Carnap's unpublished correspondence, mainly with Feigl and Hempel, is very promising and it indeed indicates some partial answers. I will concentrate on three separate points.

Carnap finished the first drafts of his intellectual autobiography in December 1956, and he sent it directly to Feigl and Hempel. He was "very dissatisfied with it" and as he said in the letter "you two are those from whom I can get the best help for the later working on."[17] As usual, Carnap was wondering about the "historical correctness" of his memory and asked his friends to think about "factual events but also about influences by books or persons on [his] conceptions or about anything else [...]" (ASP CH 11-02-10).

Hempel replied on January 14, 1957, and claimed that "it is an utterly fascinating piece, which will show [Carnap] to [his] readers from a side they never thought existed" (ASP CH 11-02-09). Hempel also suggested particular places where Carnap could shorten his autobiography. Interestingly in the light of his later remarks regarding the cutting, Hempel suggested omitting (among others) the parts about the *Jugendbewegung*.

Carnap was still bothering with the shortenings in April 1957, when his wife, Ina wrote to Hempel (who was called by his personal acquaintances as "Peter"):

> I wish, Peter, you would let me have a line taking issue with the following; if cuts are to be made (Feigl, Bohnert say: "no cuts, if possible"; I don't agree), where would your cutting emphasis lie? Mia says: on the content of publications as given by Carnap, since people can read them anyway. (Ina to Hempel, ASP RC 102-13-59.)[18]

Though at first Ina said that she "like[d] just this [suggestion] very well", after all, she had a different move in mind and did not agree with the advice that Carnap should cut the survey of his publications.[19] Instead, she claimed that "I should like

---

[17] Carnap to Hempel and Feigl, ASP CH 11-02-10. The letter is dated in Carnap's *Nachlass* as November 18, 1956 (ASP RC 091-20-18), but it was received by Hempel on December 5.

[18] The letter could be found also in Hempel's *Nachlass*, see ASP CH 11-02-07. "Mia" is Hans Reichenbach's wife, Maria Reichenbach.

[19] It would have been indeed optional for Carnap though given that Ayer's collection of *Logical Positivism* was just on his way (it was published in 1959) and he provided some fresh remarks about his recently translated papers. The University of California Press editor Robert Y. Zachary asked Quine's opinion about a possible translation of Carnap's *Aufbau* in 1961 (see Creath 1990,

to see cuts made in the more intimate material (childhood, youth movement, own children, auxiliary languages), and make it more an 'intellectual' auto."[20] What does this passage tell us?

It seems that by an intellectual autobiography Ina indeed meant an *intellectual* autobiography where the emphasis lies on the evolution of Carnap's theses, claims and results. An intellectual autobiography should not deal with the personal background and historical contexts beyond what is necessary for such a literary genre. The "intimate" parts and the personal experiences just do not add anything to the content of the philosophical claims. Presumably, it was Ina who formulated exactly this idea in a letter to Feigl:

> This morning I mailed you the first half of the autobiography. Don't be shocked about the length. Much of it is very easy reading. The part which impresses me more is always where it is not so easy reading, but then that may be my peculiarity. Before my eyes is the little book which Freud wrote as his autobiography, and which I think extremely good as an intellectual autobiography: a minimum of personal facts and a maximum about the develop[ment] and spread of the ideas, actually a history of psychoanalysis. To me this seems the ideal way of writing such a thing: people at large somehow do not seem worth the personal details, but should be fed facts primarily. Says I. (Ina/Carnap to Feigl, December 5, 1956. ASP RC 102-07-34.)

This could be justified even in the theoretical framework of Carnap and that's could be one reason why did Carnap indeed cut off the mentioned "intimate" passages. Carnap always distinguished theoretical/philosophical claims and one's attitude toward life (*Lebensgefühl*) and worldview (*Weltanschauung*). The latter underlies the former as non-theoretical complexes of cultural and social experiences along with the inherited bag of values. That was just the main point of his critic in 1932 (in the "Überwindung" article) when he claimed that though metaphysics could exhibit some positive role – namely to express one's *Lebensgefühl* – it is *not* a theoretical one.

Since the relevant passages – which contained the memories about the *Jugendbewegung* – were entitled by Carnap as "Weltanschauung: Religion, enlightenment, youth movement" he indeed seemed to identify these reflections as the background basis for his philosophy and not as *parts* of his theoretical considerations. As such, Carnap held that though the remarks about one's *Lebensgefühl* and *Weltanschauung* could be useful to understand the (often irrational or better, a-theoretical) reasons behind one's philosophy, they are useless in the evaluation of proposed arguments. In a frequently quoted passage from Carnap's intellectual autobiography – where he recalled the role and effect of Herman Nohl – he claimed that

453–454). Given that, Carnap could have known that another important book of his will be available to the English speaking world. Later on the 10th of June, 1969 Carnap wrote to Ferenc Altrichter (who was editing the selected Hungarian translations of Carnap) that "I think it will not be necessary for me to write new comments on these papers indicating my present views and how they differ from the formulations in these old papers. I made such comments at an earlier time." Listing these places Carnap did not mention his Schilpp volume. See ASP RC 027-22-01.

[20] Ina to Hempel, April 15, 1957. ASP RC 102-13-59.

> [m]y friends and I were particularly attracted by Nohl because he took a personal interest in the lives and thoughts of his students, in contrast to most of the professors in Germany at that time, and because in his seminars and in private talks he tried to give us a deeper understanding of philosophers on the basis of their attitude toward life ("*Lebensgefühl*") and their cultural background. (Carnap 1963, 4.Cf. Carnap 1957, [UCLA], Box 2, CM3, M-A3, p. B3.)

In the unpublished version, however, this passage continues with the following rarely cited remark:

> [...] since my interest was more systematically than historically oriented, I was frustrated when he [Nohl] pushed aside questions about the correctness of the views of a philosopher whose work we studied. Following his teacher Dilthey, he regarded as the main aim of philosophical study not the solution of problems, but the understanding of the ways of thinking of the various philosophers. (Carnap 1957, [UCLA], Box 2, CM3, M-A3, pp. B3-B4.)

This also seems to suggest that for Carnap, writing a mainly *intellectual* autobiography means that one should omit the elements of his *Weltanschauung*.

From this angle, we could explore the *objective meaning* of Carnap's act – though what is relevant is not the objective social configuration but the objective philosophical configuration. Carnap's act – namely to remove certain typewritten pages from a document – in a given theoretical medium become bearer of a philosophical meaning: it will be the manifestation of the idea to distinguish factual/cognitive and non-factual/non-cognitive elements. If one knows the relevant and particular philosophical stance in question and all of its commitments, then the given act of cutting the parts about worldviews (which are non-theoretical, hence non-intellectual) from an *intellectual* autobiography will be meaningful. From this point of view, it does not matter that we are talking about Carnap, Hempel, Reichenbach or any other philosopher who puts more weight on the theoretical side: what matters is that a certain act (or product) will acquire its meaning in a given objective philosophical configuration.

We could also point out that Carnap was just simply not interested in writing the autobiography and cutting the least intellectual and theoretical parts which were not known in the U.S. (as the *Jugendbewegung*) just seemed to be the most simplest move to get over the project.[21]

Carnap started to work on his autobiography in 1954; in fact, it turned out to be just a duty to him which he wanted to "avoid."[22] After all, he wrote in 1960 that "Schilpp has just sent the mss. for the Carnap-volume to the printer – perhaps it will then appear in 1961. I spent an inordinate amount of time on the writing of the 'Intellectual Autobiography', I don't do this sort of thing too well. The technical discussions are much more my sort of writing [...]."[23]

---

[21] Since every other autobiography of the Schilpp volumes starts with the author's childhood it would have been unreasonable to delete the relatively long passages about his childhood and his mother.

[22] Carnap to Feigl, June 14. 1954. ASP RC 102-08-43.

[23] Carnap to Vere Chapell, August 4, 1960. ASP RC 027-03-17.

When he was working on the intellectual autobiography, Carnap was indeed in a project with a lot of technicalities: it was the theory of probability which was his "latest and in his eyes most valuable baby" on which he was working for 30 years.[24] Due to the regular and increasingly grievous pains in his back, Carnap was concerned from time to time about the prospects of his life.[25] Since there was still a lot to do with probability, the writing and shortening of the autobiography were just a liability. In 1956, during the composition of the manuscripts, Carnap fulminated as follows: "For heaven's sake, a logician should not be asked to write a history or an autobiography, unless he is a genius like Russell!"[26] Later, in 1958, he was still quite desperate: "I am engaged in the somewhat tedious work of working over my ms. for the Schilpp volume; I shall be much relieved once this manuscript is off my hands and I can return to inductive logic."[27]

Even though Carnap had to work also on the replies, the autobiography was a "bigger chore than the former."[28] Anyways, if Carnap would have the required time, his memory would function just good as he wished, and he would have enough space for his autobiography, writing that sort of thing would have been still a huge challenge to him. In 1965, he asked Hempel to help him formulate a preface to his *Philosophical Foundations of Physics* (second edition as *Introduction to the Philosophy of Science*). He said: "You know, I am rather clumsy in formulating such things, where the non-cognitive meaning components are more important than the cognitive ones."[29] The relevant parts of the autobiography, however, were just filled with such non-cognitive passages which could not be formulated in a technical way because they formed the *basis* for all the theoretical projects of Carnap.

Again judging things from this perspective, cutting all the "intimate" parts was just the quickest move to get over the autobiography and move back to the technical projects. Carnap's correspondence documents quite well that while Schilpp insisted aggressively that he should write first the autobiography, he always tried to delay the

[24] Carnap/Ina to Hempel, August 31, 1957. ASP RC 102-13-55. According to the unpublished parts of his autobiography Carnap started to work on the questions of probability between 1941 and 1944. See Carnap 1957, [UCLA] Box 2, CM3, folder M-A5, p. P20.

[25] See for example Carnap's letter to Bochenski (October 30, 1963) where he was wondering about "how much time one has left [...]." ASP RC 027-23-40.

[26] Carnap to Hempel and Feigl, November 28, 1956. ASP CH 11-02-10.

[27] Carnap to Richard Martin, May 1, 1958, ASP RC 081-12-13. Carnap wrote to Feigl already on the 4th of February, 1955 that "the Schilpp volume is taken far more of my time than I can spare from my work on probability [...]." ASP RC 102-08-26. The same motive occurs in many other letters see e.g. ASP RC 102-08-01; ASP RC 102-07-39; but Carnap complained about it also to Hugues Leblanc, ASP RC 081-10-03.

[28] Carnap to Howard Stein, August 11, 1954. ASP RC 090-13-26.

[29] Carnap to Hempel, November 24, 1965, ASP RC 102-13-05. When Carnap wrote the preface to *Logische Syntax der Sprache*, Neurath helped him with some 'nice' formulations. See Neurath's letter to Carnap, June, 10. 1934. ASP RC 029-10-65.

task: "I have, of course, not only unconscious but also quite conscious resistance against the writing of the autobiography [...]."[30]

One could plausibly claim that omitting the "intimate" parts of the autobiography (and even some first-orderly philosophical parts) set back Carnap's historical rehabilitation for many years. Hempel was very much aware of the problem when he wrote to Ina (after she suggested cutting the personal passages):

> Would we not be shocked if there were an autobiography of Kant which had been cut down just to save on publication costs? I sympathize with the feeling and think and you should try how much Schilpp will allow; but I think that some cutting is possible where repetitions or extremely leisurely reflections occur [...]. As for Ina vs. Maria Rbch concerning where the cuts should be made if further reductions are inevitable: On the whole I would agree with Maria Rbch. I think that most of the people who are really interested in the volume will have read a good deal of Carnap's work or will be willing to look into it; and at any rate, those publications are there and available for posterity; but the material about the human side is not available elsewhere and will surely arouse a great deal of interest. (Hempel to Carnap/Ina, May 18, 1957, ASP RC 102-13-57.)

Things were settled, however, and a few years later the intellectual autobiography appeared just as we know it today. Considering things like this, it will matter that we are talking about Carnap: it was Carnap himself, who expressed with his act a certain meaning, namely the intention of avoiding to write an autobiography. It is more than just the fact that a certain theoretical/philosophical commitment surfaces in practice: a certain 'higher strata' of meaning is also *expressed* here intentionally. It is the way how we authentically grasp the conveyed meaning: "just as it was meant by the subject, just as it appeared to him when his consciousness was focused upon it" (Mannheim 1921–22/1959, 46).

The final reason – which I will just mention because it was treated in quite a detailed manner by George Reisch (2005, 2007) – is connected to the political atmosphere of the United States just before and after the Second World War. When Carnap immigrated finally to the U.S. in 1935, he found himself in a wholly different cultural and political context that he experienced earlier in Germany and later in Austria (or in Czechoslovakia). In 1935 (just before he left Europe) Carnap was about to hold a lecture tour in the U.S., especially at the New York University. Nagel was preparing the invitation and he wrote to Sidney Hook to arrange it; later Nagel quoted to Carnap some parts of Hook's letter with his own commentaries:

> »Tell Carnap that Universities throughout the U.S are becoming politically more reactionary daily and to exclude from his prospectus anything which some dumb conservative – who 'feel' these things – might regard as cultural Bolshevism. I wish I could get him to NYU for a year, but it doesn't seem possible now and we couldn't pay him enough [...].« In the light of these remarks, perhaps it would be wiser if you replaced the lecture on the relation between contemporary philosophy and culture by something less full of dynamite. (Nagel to Carnap, January 5, 1935. ASP RC 029-05-16.)[31]

[30] Carnap to Feigl, November, 14. 1955. ASP RC 102-08-06. On Schilpp's forceful letters see the correspondence of Carnap and Feigl, ASP RC 102-08-07 and ASP RC 102-08-09.

[31] In turn Ina replied that "[t]hough the chance of a longer stay there is very little, Carnap is glad that H. knows about him and is also grateful for the advice not to show apparently that he is a

The situation just got worse after the Second World War under the McCarthy-area and so Carnap had to rethink and stow his European socialist and political sensitivity away. As Reisch (2005) showed in his sociology of science book with great clarity, there was only one way to uphold Carnap's admitted professionalism and significance: he had to fall back to the "icy slopes of logic" (Carnap et al. 1929/1973, 317).

If even the mentioning of the relation between culture and philosophy by an allegedly East-European socialist in the mid-1930s was just so dangerous and triggered the concept of "cultural Bolshevism", some parts of his intellectual autobiography could cause a certain philosophical and cultural trauma in the philosophy departments.[32] The earlier phases of the a-political *Jugendbewegung* and the later political phases was uniquely German and considering the fact some members of the movement ended up in either communist or Nazi groups could not help the carefully constructed politically-neutral picture of Carnap. From *this* perspective, cutting the "intimate" parts of the autobiography was perhaps the right move to shorten the manuscripts.

Though Carnap did not even mention or referred to the political atmosphere of the United States, or that he had in mind such reasons to cut the "intimate parts", at this point we are facing the third "strata of meaning", i.e. the documentary-meaning in the Mannheimean sense. Even if Carnap's act points to theoretical commitments, and even if he evidently expressed his disinterest in the autobiography, cutting the politically (possibly) sensitive parts 'unintentionally' *documents* certain trends and the socio-political environment of the 1950s and 1960s.

As Peter Galison (1996, 35) wrote, "[...] people move across oceans with relative ease, complexes of ideas do not." If Carnap would and could publish his intellectual autobiography as he first imagined and wrote it (i.e. if he could 'move' his own *Jugendbeweger* past officially), then perhaps he could have helped the understanding of his ideas and that socio-cultural environment which gave rise to his informal and even technical philosophy.

Acknowledgement I would like to thank Christian Damböck and Thomas Uebel that they read the manuscript and provided helpful comments. I am indebted to the Carnap Archives at Los Angeles (Rudolf Carnap papers (Collection 1029). UCLA Library Special Collections, Charles E. Young Research Library) and at Pittsburgh (Rudolf Carnap Papers, 1905–1970, ASP.1974.01, Special Collections Department, University of Pittsburgh), and to the Hempel Archive also at Pittsburgh (Carl Gustav Hempel Papers, 1903–1997, ASP.1999.01, Archives of Scientific Philosophy, Special Collections Department, University of Pittsburgh.) for the permission to quote the archive materials. All rights reserved. I cite the Rudolf Carnap and Carl Hempel Archives from Pittsburgh as follows: ASP RC XX-YY-ZZ and ASP CH XX-YY-ZZ, where XX is the box number, YY the folder number, and ZZ the item number; the UCLA archive as Carnap 1957, [UCLA] fol-

'Cultur Bolshevist'. But it seems not necessary to replace the lecture-title on the relation between contemporary philosophy and culture by an other, because Carnap may speak about anything under this vague title." Carnap/Ina to Nagel, January 23, 1935. ASP RC 029-05-15. Hook had a quite complex relation to the logical empiricists, see Reisch (2005).

[32] In his personal correspondence Carnap complained a lot about the atmosphere and attitude both of his department at Chicago and about other philosophy departments in the U.S.

lowed by box, folder, and page numbers. Normal underlining in the quotations is made with pencil in the archive materials. The research was supported by the Hungarian National Grant of Excellence.

Bibliography

Aufmuth, Ulrich. 1979. *Die deutsche Wandervogelbewegung unter soziologischem Aspekt.* Göttingen: Vandenhoek und Ruprecht.

Bias-Engels, Sigrid. 1988. *Zwischen Wandervogel und Wissenschaft: Zur Geschichte von Jugendbewegung und Studentenschaft 1896–1920.* Cologne: Verlag Wissenschaft und Politik.

Bouveresse, Jacques. 2012. Rudolf Carnap and the Legacy of *Aufklärung*. In *Carnap's Ideal of Explication and Naturalism*, ed. Pierre Wagner, 47–62. Palgrave: Macmillan.

Carnap, Rudolf. 1934/1937. *Logical Syntax of Language*. Trans. A. Smeathon. London: Kegan Paul Trench.

———. 1963. Intellectual Autobiography. In *The Philosophy of Rudolf Carnap*, ed. Paul A. Schilpp, 3–84. Chicago/LaSalle: Open Court.

Carnap, Rudolf, Hans Hahn, and Otto Neurath. 1929/1973. Scientific Conception of the World: The Vienna Circle. In *Empiricism and Sociology*, ed. Marie Neurath and Robert S. Cohen, 299–318. Dordrecht: Reidel.

Carus, André. 2007a. *Carnap in Twentieth-Century Thought: Explication as Enlightenment.* Cambridge: Cambridge University Press.

———. 2007b. Carnap's Intellectual Development. In *The Cambridge Companion to Carnap*, ed. Michael Friedman and Richard Creath, 19–42. Cambridge: Cambridge University Press.

Creath, Richard, ed. 1990. *Dear Carnap, Dear Van: The Quine-Carnap Correspondence and Related Work.* Berkeley: University of California Press.

———. 2007. Vienna, the City of Quine's Dreams. In *The Cambridge Companion to Logical Empiricism*, ed. Alan Richardson and Thomas Uebel, 332–345. Cambridge University Press: Cambridge.

Dahms, Hans-Joachim. 2016. Carnap's Early Conception of a 'System of all Concepts': The Importance of Wilhelm Ostwald. In *Influences on the Aufbau*, ed. Christian Damböck. Dordrecht: Springer.

Damböck, Christian. 2012. Rudolf Carnap and Wilhelm Dilthey: 'German' Empiricism in the *Aufbau*. In *Rudolf Carnap and the Legacy of Logical Empiricism*, ed. Richard Creath, 67–88. Dordrecht: Springer.

Demeter, Tamas. 2012. *Weltanschauung* as a Priori: Sociology of Knowledge from a 'Romantic' Stance. *Studies in East European Thought* 64: 39–52.

Dilthey, Wilhelm. 1968. *Weltanschauungslehre. Abhandlungen zur Philosophie der Philosophie,* Gesammelte Schriften, Vol. 8, 4th ed. Stuttgart/Göttingen: B.G. Tuebner/Vandenhoeck & Ruprecht.

Flitner, Wilhelm 1986. *Erinnerungen 1889–1945*, Gesammelte Schriften, Bd. 11. Paderborn/München/Wien/Zürich: Ferdinand Schöningh.

Frank, Philipp. 1949. Introduction – Historical Background. In *Modern Science and its Philosophy*, 1–52. New York: George Braziller.

Gabriel, Gottfried. 2004. Introduction: Carnap Brought Home. In *Carnap Brought Home – The View from Jena*, ed. Steve Awodey and Carsten Klein, 3–23. Chicago/LaSalle: Open Court Publishing.

Galison, Peter. 1996. Constructing Modernism: The Cultural Location of *Aufbau*. In *Origins of Logical Empiricism*, ed. Ronald Giere and Alan Richardson, 17–44. Minnesota: University of Minnesota Press.

Haller, Rudolf, and Heiner Rutte. 1977. Gespräch mit Heinrich Neider: Persönliche Erinnerungen an den Wiener Kreis. *Conceptus* 1: 21–42.

Jeffrey, Richard. 1994. Carnap's Voluntarism. In *Logic, Methodology, and Philosophy of Science IX*, ed. Brian Skyrms DagPrawitz and Dag Westerståhl, 847–866. Elsevier: Amsterdam.

Kamlah, Andreas. 2013. Everybody Has the Right to Do What He Wants: Hans Reichenbach's Volitionism and Its Historical Roots. In *The Berlin Group and the Philosophy of Logical Empiricism*, ed. Nikolay Milkov and Volker Peckhaus, 151–175. Dordrecht: Springer.

Landauer, Carl. 1978. Memories of Hans Reichenbach. University Student: Carl Landauer. In *Hans Reichenbach: Selected Writings: 1909–1953*, ed. Maria Reichenbach and Robert S. Cohen, vol. 1, 25–31. Dordrecht: Reidel.

Laquer, Walter Z. 1962. *Young Germany. A History of the German Youth Movement*. New York: Basic Book Publishing Co., Inc.

Mannheim, Karl. 1921–22/1959. On the Interpretation of *Weltanschauung*. In *Essays on the Sociology of Knowledge*, ed. Paul Kecskemeti, 33–83. London: Routledge/Kegan Paul.

Messer, August. 1924. *Die freideutsche Jugendbewegung: Ihr Verlauf von 1913 bis 1922*. Langensalza: Beyer.

Quine, Willard van Orman. 1971. Homage to Rudolf Carnap. In *PSA 1970. In Memory of Rudolf Carnap*, ed. Roger C. Buck and Robert S. Cohen, xxii–xxxv. Dordrecht: D. Reidel Publishing.

Reisch, George. 2005. How the Cold War Transformed Philosophy of Science. In *To the Icy Slopes of Logic*. Cambridge: Cambridge University Press.

———. 2007. From 'the Life of the Present' to the 'Icy Slopes of Logic': Logical Empiricism, the Unity of Science Movement, and the Cold War. In *The Cambridge Companion to Logical Empiricism*, ed. Alan Richardson and Thomas Uebel, 58–87. Cambridge University Press: Cambridge.

Tuboly, Adam Tamas. forthcoming. From the Jugendbewegung to the Vienna Circle: Carnap's Cooperation with Hans Freyer. Forthcoming manuscript.

Uebel, Thomas. 2003. Philipp Frank's History of the Vienna Circle: A Programmatic Retrospective. In *Logical Empiricism in North America*, ed. Gary Hardcastle and Alan Richardson, 149–169. Minneapolis: University of Minnesota Press.

Werner, Meike G. 2003. Moderne in der Provinz. In *Kulturelle Experimente im Fin de Siècle Jena*. Göttingen: Wallstein.

Reviews

Günther Sandner, *Otto Neurath. Eine politische Biographie*. Vienna: Paul Zsolnay Verlag, 2014

Otto Neurath—who needs no introduction to readers of this Yearbook—lived a life that was anything but uneventful. Moreover, it was played out in far more than one professional arena. The biography under review by Günther Sandner is not the first. Besides his son Paul Neurath's highly informative biographical sketch that observed disciplinary neutrality, we have to date Karola Fleck's early study that focused upon the making of the philosopher and two more recent ones that tell his intellectual biography in terms of a series of interdisciplinary foundational debates, but again with an eye primarily to the philosopher.[1] Sandner's highly readable biography differs from these, first of all, in being of monograph length and, second, in focusing upon Neurath as a political thinker—that is, "political" in rather general terms. This affords the advantage of considering aspects of Neurath's life and activity that found little discussion in his philosophy as such, but nevertheless were formative early on or decisive for his later development. Therein lies the strength of this book. And since Neurath was, without doubt, the most political animal among the logical positivists, whatever we can learn here will also inform our view of this expressly holistic thinker as a philosopher. But those conclusions are left for the reader to draw.

---

[1] Paul Neurath, "Otto Neurath (1882–1945). Leben und Werk" in P. Neurath, E, Nemeth (eds.), Otto Neurath oder die Einheit von Wissenschaft und Gesellschaft, Vienna: Böhlau, 1994, 13–96. (A shorter English version is "Otto Neurath (1882–1945). Life and Work" in E. Nemeth and F. Stadler (eds.), *Encyclopedia and Utopia. The Life and Work of Otto Neurath*, Dordrecht: Kluwer, 1996, 15–28.) See also Karola Fleck, Otto Neurath. *Eine biographisch-systematische Untersuchung*. University of Graz, 1979, rev. and trans. in N. Cartwright, J. Cat, K. Fleck, T. Uebel, *Otto Neurath. Philosophy Between Science and Politics*, Cambridge: Cambridge University Press, 1996, 7–88; T. Uebel, "Otto Neurath—Leben und Werk", in *Internationale Bibliographie zur österreichischen Philos*ophie Bd. 10: 1991/1992, Amsterdam: Rodopi, 2005, 7–52, and J. Cat, "Otto Neurath", in E. Zalta (ed.), *The Stanford Encyclopedia of Philosophy*, internet, 2011 (http://plato.stanford.edu/archives/win2011/entries/neurath).

F. Stadler (ed.), *Integrated History and Philosophy of Science*, Vienna Circle Institute Yearbook 20, DOI 10.1007/978-3-319-53258-5

Consider "The Young Neurath" (Ch. 1). Apart from presenting the roles of his father Wilhelm Neurath and of the sociologist Ferdinand Tönnies and from allowing us to engage with three very expressive photographs—one as a bright and prepossessing schoolboy, one as an optimistic youth in a group with a happy Popper-Lynkeus (one of his heroes) and his father Wilhelm looking on quizzically (a truly prophetic picture), and one with his first wife Anna Schapire and her sister Rosa (who was to become a well-known art historian)—here we learn of an element of Neurath's world which on reflection may readily appear presupposed but was rarely if ever topicalized as such. That is his engagement with turn-of-the-century feminism, first in correspondence with the then popular if controversial Swedish author Ellen Key, then with the work and person of his wife Anna who published on the history of the women's movement and the ongoing struggle for women's rights in the work place. Anna was responsible, so Sandner, for impressing upon Neurath the importance of the so-called women's question. Another interesting issue raised by Sandner is the role that his Jewish identity may have played in his early life (as opposed to later on). Where some commentators pointed out that in many ways Neurath's background was typical for children of the first generation of emancipated Austro-Hungarian Jews, Sandner points out that his father had converted to Catholicism, his mother was a German-born Protestant and Otto himself had been baptized according to Catholic rites. The apparent blindness to anti-semitic passages in Marlowe's *Faust* which Neurath edited in 1906 certainly contrasts with the awareness of racial stereotyping that in the years of his English exile in the 1940s he planned to analyze historically in a book-length study of toleration (but never completed).

Chapter 2, "The Teacher of War Economics", turns to Neurath's pre-World War I academic career. His work at the New Viennese Commercial Academy from 1907–1914 laid the basis for his life-long interest in civic education (in economics, political and social science) attested also by numerous publications at the time. Meanwhile war economics, the study of the transformation of national economic life under war conditions, became his professional specialization. Here we learn previously unknown details about Neurath's service during World War I in the Austrian War Ministry as a group leader of its "Scientific Committee" (with the post-war Austrian Foreign Secretary Otto Bauer working under him) that was collecting and systematizing statistical material, and about a somewhat parallel effort towards the foundation of a Museum for War Economics in Leipzig in 1918. Neurath's concerns in the Vienna ministry and the Leipzig museum were of a scholarly rather than nationalist nature and so far his party-political position remained indeterminate. From 1908 Neurath had been writing for the periodical *Kunstwart* (which was distinctly anti-socialist before the war, unlike after the German revolution when Neurath used it as a forum for his socializations plans) and in 1917 he was an invited speaker in one of the cultural congresses at Burg Lauenstein addressing alongside Max Weber the leadership of the sometimes politically ambivalent German Youth Movement (to the left wing of which belonged his later philosophical colleagues Carnap and Reichenbach). By 1919, Neurath had joined the Social Democratic Party and took up a highly exposed (which he described in retrospect as merely administrative) in the Bavarian revolution as chief of its central economic office overseeing its socialisation efforts. In consequence his accreditation as university teacher in Heidelberg, awarded only 2 years previously, was withdrawn—ending his academic career.

Chapter 3, "Vita Activa: Total Socialization", deals with the ideas and plans that propelled Neurath into public consciousness: his advocacy, promulgated by numerous articles and pamphlets, of efforts to organize the national economy by central planning according to the principle of production for social need rather than capitalist profit. Sandner provides general background, outlines Neurath's scheme, describes the stages that led to his appointment in Munich and the subsequent post-revolutionary imprisonment, and notes both the democratic potential of Neurath's plans and the well-known criticisms of their practical shortcomings. What's new here concerns mainly the protracted negotiations that led to Neurath's extradition to Austria (allowing him to avoid the lengthy jail sentence he had received), his brief relations with the industrialist, political author and later German Foreign Secretary Walter Rathenau (murdered by anti-Semites in 1922), and the circumstances of his production of a socialization pamphlet for a "Jewish planned economy in Palestine". While the analysis of Neurath's socialization plans generally and of the criticisms they received does not go far enough to allow readers to see why they differ significantly from the planned economies that history did, after all, bequeath on Eastern Europe and Asia, Sandner is correct to note this.

In Chap. 4 we find Neurath back in the now republican Vienna, ruled by the Social Democrats who retained the city administration into the 1930s although the rest of Austria already in the early 1920s fell back to conservative parties (putting an end even to the modest Austrian efforts at socialization). Besides his commitment to workers' education throughout the period, Neurath's involvements reached from the short-lived Institute for Communal Economy and the urban settlement movement in the early 1920s to his role in the Austrian Werkbund (an association of artists, architects and designers) in the later 1920s and CIAM's 1933 architects' congress in Athens. Most important, however, were the founding of his innovative Social and Economic Museum with its internationally applauded method of pictorial statistics and, last but not least, his role in the Vienna Circle. Sandner largely confirms what is already known but adds numerous personal and institutional details that provide local color and so successfully conveys the dense networks created by Neurath.

Chapter 5 considers Neurath in emigration: first the new start of his museum's picture-statistical work in Holland and its expansion to the USA and the organizational work for the unity-of-science congresses from 1934 to 1939 and the International Encyclopedia project with University of Chicago Press, then his second flight in a small fishing boat to England and his restart of his work there with Marie Reidemeister (soon to become his third wife) and with the documentary film-maker Paul Rotha. Neurath was much concerned with the question of the re-education of the population of the soon to be defeated German Reich and he participated in the debates about the democratic standing of Plato's conceptions of the state (discussions that had been prompted long before Popper, we may add, by the Oxford don and later Labour politician Richard Crossman). Not without interest is Sandner's finding that Neurath intentionally kept his distance from the political infighting amongst the Austrian emigrées.

What then, finally, about Neurath's politics? Many remain puzzled by the picture-statistical work he and his team did in Moscow from 1931 to 1934 for the publication of the results of the first 5-year plan and of projections for the second one. Sandner writes that Neurath's view of the Soviet experiment in communism, which in the early 1920s he had regarded "not without sympathy" (227), markedly "deteriorated" (232) in light of his experiences there. What had attracted him originally was the opportunity to participate in an actual attempt to refashion a national economy—which he was led to regard as far more successful than it was—but throughout his engagement in Moscow (which was sanctioned by the Austrian Social Democrats who at home refused cooperation with communists) he stuck to his own ideological position as a democratic socialist which there was much disparaged. (Criticisms of the Soviet Union which Neurath voiced later alienated former friends like the architect Schütte-Lihotzky). His picture-statistical involvement apart, Neurath's position does not therefore seem to have differed much from that of many on the left in Western Europe in the early 1930s.

It is in his 1928 monograph partially translated as "Personal Life and Class Struggle" that Sandner locates the "key text for understanding the political Neurath" (211).[2] Sandner is undoubtedly right to stress as central Neurath's understanding of Marxism as "social Epicureanism", in contrast both to then current attempts to re-Hegelianize it (after its previous mechanization, as it were, by the German party orthodoxy), and to relate this understanding directly to Neurath's theory of life conditions which provided the systematic foundation of Neurath's socialization plans (as in his 1925 monograph *Economic Plan and Calculation in Kind*).[3] It may be doubted, however, whether already by 1937 Neurath had given up on his hope for a socialist future, as Sandner tentatively suggests with regard to a paper published as part of an ill-fated attempt at cooperation with Horkheimer's Frankfurt Institute of Social Research. Further discussion of the political dimension of this particular episode (the overall dynamics of which Sandner judges correctly) would have provided, one suspects, a welcome opportunity to deepen our understanding of Neurath's take on Marxism up to then. By 1945, in any case, Neurath did distance himself from Marxism, as Sandner documents with reference to a letter to Josef Frank (the architect brother of his Vienna Circle colleague Philipp). In a manuscript from the same year that remained unpublished in his lifetime he summarized his later pluralist views on politics, education and scientific intervention in social processes.[4] In his biographer's words: "Democracy meant for him the communication between experts and citizens on the same level. It meant the search, on the basis of

---

[2] See Neurath, *Empiricism and Sociology* (ed. by M. Neurath and R.S. Cohen), Dordrecht: Reidel, 1973, 249–298.

[3] See Neurath, *Economic Writings* (ed. by T. Uebel and R.S. Cohen), Dordrecht: Kluwer, 2004, 405–465.

[4] See Neurath, "Visual Education. Humanisation versus Popularisation" (ed. by J. Manninen), in Nemeth and Stadler, *op. cit.*, 245–335.

knowledge common to all, for a compromise that is acceptable to all affected. A decision of the majority against the minority represents only an exception, but not rule of democratic processes." This is hardly a political *theory*, to be sure, but,as Sandner adds, this is "a substantive *approach* to the theory of democracy incorporating deliberative and participatory elements" (292, emphasis added). And he notes that in this letter to Frank Neurath admitted that the tensions between a central organization of the economy and individual freedom are not easily resolved—and had not been so to date.

Neurath's life and work, Sandner suggests, was not only "a product of his time" but also a "formative influence on his time", propelled by a philosophical-political program that "still today remains worthy of discussion" (297). His articulation of Neurath's legacy focuses on the "democratization of knowledge" much-discussed nowadays and on the contribution his picture-statistical methods of communicating social facts can make to this. Sandner acknowledges Neurath's mistakes concerning the separation of the science of political economy from politics itself, but notes that despite his failed prognoses in mid-life about the imminent advent of socialism in the West he also would never have regarded the end of state socialism as the "end of history". Neurath's criticism of undirected—certainly unchecked—capitalist expansion, irrespective of the limits of the planet or the needs of its population, finally, is judged justified more than ever today. In sum, Sandner's biography admirably succeeds in making accessible Neurath's life work across several disciplines by relating his seemingly disparate efforts to the humanistic utopianism at their core and provides a balanced assessment of his achievements.

Thomas Uebel
(University of Manchester)

Eino Kaila, *Human Knowledge: A Classic Statement of Logical Empiricism*. Translated by Anssi Korhonen; edited by Juha Manninen, Ilkka Niiniluoto, and George A. Reisch. Chicago, Illinois: Open Court, 2014, pp. xxvii + 217

Eino Kaila (1890–1958) is one of the less well-known figures within the logical empiricist movement. Although there has been some research on his philosophical work in recent years, Kaila's contribution to the logical empiricist project is still in need of closer examination. The present volume should prove as an excellent basis in this respect. In fact, Kaila's *Inhimillinen tieto* is a classic of early/mid twentieth-century philosophy of science. It is therefore all the more important that the book is now available in the translation by Anssi Korhonen.

The volume comprises ten chapters and an introduction by the editors Juha Manninen and Ilkka Niiniluoto. As the editors point out in their introduction, Kaila conceived of his book (published in the Finnish original in 1939) both as a textbook of scientific philosophy for laymen and as a systematic introduction to logical empiricism for professionals. Rudolf Carnap, with whom Kaila stood in close contact, welcomed Kaila's contribution especially for its paying attention to the "historical connections," since these, as Carnap admitted, were "mostly ignored" in the existing publications by the logical empiricists. Furthermore, Carnap suggested to publish the book in English translation for the *Library of Unified Science* which in turn was published, in the Dutch exile, by Otto Neurath. However, nothing came of Carnap's suggestion because Holland was soon occupied by German troops, Neurath had to flee to England, and Finland went to war against the Soviet Union in the summer of 1941. But how came it that Kaila had such splendid connections to the members of the Vienna Circle? Here, it must be seen that academic philosophy in Finland had for a long time been dominated by Hegelian idealism which, according to the editors, was represented by the "national philosopher" Johan Vilhelm Snellman (1806–1881). Kaila, even in his early years, was not at all attracted by idealism. Rather, he engaged in the philosophy of science, focusing on Machian positivism and its rejection of atomism. Kaila himself defended the reality of atoms and argued for their being part of a "mind-independent causal nexus." In 1926, he published his monograph *Die Prinzipien der Wahrscheinlichkeitslogik*, where he critically discussed both the views of probability of Edgar Zilsel and Hans Reichenbach. Herbert Feigl, in his 1927 dissertation *Zufall und Gesetz* (which was supervised by Moritz Schlick), critically evaluated Kaila's monograph. In 1929, Kaila (on invitation by Schlick) decided to visit Vienna in order to participate at some of the Vienna Circle's meetings. In the Circle's 1929 manifesto "Wissenschaftliche Weltauffassung – Der Wiener Kreis," Kaila was mentioned as one of the thinkers close to the "scientific world-conception" of the Circle. This had to do in the first place with Kaila's methodological orientation which implied that there is no sharp difference between philosophy and special scientific disciplines and that philosophy itself should proceed by making use of exact methods. However, as concerns questions of systematic outlook, Kaila, like Reichenbach, defended some sort of probabilistic realism, particularly against Carnap's declaration that the realism controversy is meaningless.

This brings us to the book's ten chapters which are tied together by one red thread, namely (the unifying idea of) *invariance*. More precisely, the book is subdivided in three parts. Part One deals with the problem of theory formation, Part Two with the formal truth of theories, and Part Three with the empirical truth of theories. Invariance plays an essential role in all three parts. However, it is especially Part One where Kaila develops his invariantist approach to science and scientific theory construction. As he declares in the preface, for him "the logical empiricist conception of knowledge is the culmination of two and a half millennia of development in human ideas" (xxvi). Yet, it must be seen that Kaila, by invoking invariance, contributed an own and very specific version of the logical empiricist conception of knowledge. Heavily inspired by Ernst Cassirer's *Substanzbegriff und Funktionsbegriff* (1910), Kaila characterized the aim of science as the "search for invariances" (3). By 'invariance' he meant something like regularity, or lawfulness; but he also meant by it the stability, or constancy, of physical entities like energy. On the whole, it is invariances which, according to Kaila, are the object of both scientific and prescientific knowledge. Or, as he puts it at the beginning of Chap. 1: "As the invariances that we discover are more general, the more we succeed in satisfying our pursuit of knowledge." (ibid.) Thus there is a hierarchy of invariant systems ranging from everyday perceptual objects to the most stable and lawful objects of science. The outstanding characteristic of the objects of science is that they are *idealized.* According to Kaila, in science "we *round off* everything in thought" (10), that is, we "*rationalize* our concepts – for instance the concept of acceleration – to give them that exactness, precision, and simplicity that is not possessed by the corresponding phenomena of experience" (ibid.). Nevertheless, the search for invariances leads, according to Kaila, to substantive knowledge. Although it is not perceptual qualities which are grasped by scientific knowledge, we are in position to acquire knowledge of certain *structural features* which, in mathematized science, usually have the status of *isomorphisms*. It is for this reason that Kaila thinks that "it is wrong to say that we know *nothing* of things-in-themselves; after all we know their structure" (14).

Chapters 2, 3 and 4 reconstruct the historical development from the Greeks up to Galileo, Newton, and Leibniz. For Kaila, Galileo is the hero of this story. For it was Galileo who brought together the two decisive components of scientific knowledge: the search for invariances, on the one hand, and the requirement of verification, on the other. Whereas Aristotle raised the question '*What?*' and accordingly looked for the substance, or essence, of things, Galileo raised the question '*How?*' and accordingly looked for functions, or as Kaila alternatively puts it, "*relational invariances*" (51). Questions about essences were completely ignored by Galileo, which in turn, in Kaila's eyes, makes him "one of the forerunners of logical empiricism" (53). However, with Descartes, the empiricist impetus was rudely stopped: "This distinguished mathematician, despite being given the honorific title 'father of modern philosophy', was far behind Galileo in his conception of knowledge. In Galileo we find a fruitful balance between the search for invariances and the requirement of empirical verifiability. But with Descartes this balance tilts toward Plato and a postulate of invariance. Empirical verifiability, it is suggested, is not necessary in principle, for we are supposed to know the laws of nature in advance." (59) Unlike Galileo, Descartes, by raising the question '*Why*,' was looking for 'ultimate causes' and thereby stepping back to Aristotelian essentialism. But then, Kaila rather dramatically declares, the "gigantic figure of Newton" (61) entered the stage. With Newton, the Galilean conception of knowledge got saved, that is, according to Kaila "Newton redirects the course of modern science, rescuing it at a moment when Cartesianism was leading it away from the right path" (61). By rejecting *a priori* speculative hypotheses about the essences of phenomena and their causes, Newton returned to the empirical basis of science. As early as in his *New Theory of Light and Colours* of 1671, Newton refused to answer Aristotelian and Cartesian *what-* or *why*-questions: "Science has no other task than to start from experience and state the exact laws of phenomena that will help other phenomena to be predicted. That famous slogan, 'Hypotheses non fingo,' is already presupposed in this first work." (62) With Leibniz, this whole development reaches its culmination. For, according to Kaila, it was Leibniz who, in terms of his "*principe de l'observabilité*, most forcefully articulated the requirement of empirical verifiability. Thus, like Galileo and Newton, Leibniz – the alleged "radical rationalist" (67) – should be seen as a forerunner of the modern, i. e. logical empiricist, conception of knowledge.

Chapter 5 closes Part One by reflecting on the problem of induction and its relation to the concept of probability. As Kaila briefly indicates, the task of an 'inductive logic' in his view is illusory. For him, the probability that we assign to inductive generalizations is purely psychological. It has to do exclusively with the "way of discovery" (82), whereas logic is restricted to the "way of demonstration" (ibid.). Accordingly, an inductive *logic* would be a *contradictio in adiecto.*

Part Two of the book is subdivided in two chapters. Chapter 6 deals with logical truth, Chap. 7 with mathematical truth. As concerns logical truth, Kaila gives an instructive and very readable overview over the basic elements of modern first-order logic. He thereby draws on results provided by David Hilbert, Bertrand Russell, Ludwig Wittgenstein, and Alfred Tarski. Furthermore, he addresses Kurt Gödel's work on the so-called decision problem and finally concludes that logical truths are "consequences of definitions" (120) and are therefore to be seen as analytical sentences. Interestingly enough, Kaila in this context anticipates certain ideas by W. V. O. Quine, claiming that "the analyticity and syntheticity of a sentence is a 'relative matter' that depends on how certain concepts have been defined" (116). As concerns *mathematical* truth, Kaila, at the end of Chap. 7, introduces what he calls "the first main thesis of logical empiricism" (136). What this thesis says is that the metalogical statements 'Statement L is analytic' and 'Statement L is *a priori*' are equivalent. The so-called second main thesis of logical empiricism says that every statement concerning reality must have real content. This in turn comes very close to what Carnap (in his "Testability and Meaning") called the Principle of Testability. Kaila concludes Part Two by claiming that "Kant's basic question, 'How are synthetic *a priori* statements possible?' is a mistake because there are no such statements." (140) On Kaila's own account only analytic statements are *a priori* and *vice versa.* Synthetic statements, on the other hand, are *a posteriori*, i.e., dependent exclusively on experience.

Part Three, which deals with empirical truth, is focused on such synthetic statements. At its very beginning, in Chap. 8, Kaila introduces the so-called third main thesis of logical empiricism, namely the Principle of Translatability which says that every theory (or set of theoretical statements) must be translatable into the language of experience. However, Kaila qualifies this principle by conceding that not every factual statement must be capable of a definitive verification (or falsification). He thereby criticizes the "radical positivist" (147) positions of Ludwig Wittgenstein and (especially) Moritz Schlick who, in his view, required that every factual statement be translatable to statements concerning 'the given.' Yet, in the further development of logical empiricism this radical view became liberalized by the weaker requirement of testability. As Kaila further points out, there is no empirical statement which is immune against revision. On the other hand, he goes not so far as to defend some sort of 'coherence theory of knowledge,' albeit "some extremists among the logical empiricists" (156), especially Otto Neurath, argued in favor of such a theory. On the whole, it remains somewhat unclear what Kaila's own position in this context amounts to. The best guess seems to be that he intends to defend some sort of Duhemian 'holism,' as regards the relation of theory and experience. At any rate, Kaila explicitly states that "[w]e must [...] give the principle of testability a broad interpretation, so that a theory in its entirety can be regarded as 'one sentence'" (170). Furthermore, Kaila rejects all forms of metaphysics, understanding by 'metaphysical' a sentence which is intended as a factual sentence but does not have any experiential consequences. He directly criticizes Heidegger's "essentialism" and "existentialism" and banishes it (in an overtly Carnapian manner) from the area of philosophy as "something like a lyrical outburst" (173). Chapter 9 deals with the "logic of physical theories." It contains an interesting interpretation – and justification – of "micro-physical" theories. In Kaila's view, "a sentence of a physical theory cannot be ruled out as 'metaphysics' solely on the grounds that it fails to depict any specific phenomenon of experience" (195). Rather, "[f]rom a logical point of view, there is nothing wrong with developing a micro-physical theory as far beyond the 'threshold of observation' as one may wish, in which case the theory will of necessity contain many sentences that cannot be tested in experience, as long as they are considered in themselves" (ibid.). Again, Duhemian holism, drives Kaila's argumentation, thus anticipating Quine anew. The concluding Chap. 10 is devoted to what Kaila calls "logical behaviorism." By 'logical behaviorism' he means the articulation of the following, so-called fourth main thesis of logical empiricism: Sentences about a subject's immediate experience are equivalent to certain sentences about the states in the subject's body. Simply put, Kaila in this connection recapitulates Carnap's conception of the notorious mind-body problem. His position seems to be that of a 'moderate physicalist.' However, how the questions pertaining to the mind-body problem are to be answered is, according to Kaila, "for future experience to decide" (205).

Given the increasing interest in Kaila's variant of logical empiricism, the present volume is a valuable source for scholars interested in the history of philosophy of science. Moreover, *Human Knowledge* deserves to be recommended to those who want to get a systematic overview over the principal tenets, claims and arguments of the logical empiricist project.

Matthias Neuber
(Tübingen)

Ernst Mach Studienausgabe: *Band 1: Analyse der Empfindungen* (hg. v. Gereon Wolters), Xenomoi Verlag, Berlin, 2008; *Band 2: Erkenntnis und Irrtum* (hg. v. Elisabeth Nemeth u. Friedrich Stadler), Xenomoi Verlag, Berlin, 2011; *Band 3: Die Mechanik in ihrer Entwicklung. Historisch-kritisch dargestellt* (hg. v. Gereon Wolters u. Giora Hon), Xenomoi Verlag, Berlin, 2012; *Band 4: Populär-Wissenschaftliche Vorlesungen* (hg. v. Elisabeth Nemeth u. Friedrich Stadler), Xenomoi Verlag, Berlin, 2014

Readers of this Yearbook need not be told of Ernst Mach's towering importance in the history of philosophy of science nor be reminded of his vast if controversial legacy. Giving notice here of the publication of the first three volumes of the *Ernst Mach Studienausgabe*—its first third—can amount, it might seem, to no more than giving notice of new bottles in which a famous old vintage is now available. Such notices can be useful in pointing fellow professionals to new teaching resources and directing fellow enthusiasts to purchases for personal consumption or for presents to those with a youthful interest in the subject matter. On this occasion, however, it must be stressed as well that the re-bottling has been expertly done.

In handsomely produced sturdy paperbacks we now find readily available *Band 1: Analyse der Empfindungen* (AdE, 1886), *Band 2: Erkenntnis und Irrtum* (EuI, 1905), and *Band 3: Die Mechanik in ihrer Entwicklung. Historisch-kritisch dargestellt* (MiiE, 1883), *Band 4: Populär-Wissenschaftliche Vorlesungen* (PWV, 1895). The numerous original steel engravings and illustrations, for instance, were digitalized individually for better reproduction, avoiding common failings of less careful republications. The texts too have been newly set—helpfully so with indications of original page breaks—according to the last edition authorized by Mach himself. Thus we get *AdE* in the form of the sixth edition of 1911, *EuI* in that of the second edition of 1906, *MiiE* in that of the seventh of 1912, and PWV in that of the fifth of 1923. Given the changes wrought by Ludwig Mach under the guise of supposedly verbal instructions for still later editions by his late father, this policy (followed for the entire *Studienausgabe*) is very well justified.

Each volume editor or team of volume editors—*Band 1*: Gereon Wolters, *Band 2 and 4:* Elisabeth Nemeth and Friedrich Stadler, *Band 3*: Gereon Wolters and Giora Hon—has performed tasks that become increasingly important as Mach's present recedes into our past. Not only have Mach's own sometimes incomplete or misleading bibliographic references in the original been corrected throughout, but numerous explanatory footnotes have been added. Clearly marked as such but integrated among the original footnotes below the relevant text, these notes serve a variety of purposes. Some explain what are now uncommon phrases of Mach's or what were then and now domain-specific concepts; others give biographical detail of the very many scientists mentioned by Mach often only in passing (many of them largely forgotten today except by specialist historians); others still provide further elucidation of some of Mach's sometimes overly succinct descriptions of the phenomena at issue or of related ones. With these editorial notes Mach's texts gain much in accessibility not only for first-time readers; indeed, re-reading them becomes a surprisingly refreshing experience.

Each volume also has an individual Preface by its editor/s which gives the background of Mach's intellectual biography against which the book should be read, outlines its publication history and presents the book's central thesis without undue scholastic clutter to stimulate the reader's interest.

Thus the Preface to *Band 1* wastes no time in on arguing against a misinterpretation that Mach's "home-spun" epistemology of science in *AdE* has often been subjected to—that of being a phenomenalist idealist—but instead sets out right away his "methodological naturalism" which regards human knowledge as best understood by the categories of the most advanced natural sciences that pertain directly to the subject (then it was sense-physiology to which Mach himself had made significant original contributions). Moreover, what may appear as mere sensationalist reductionism on the part of Mach is swiftly revealed as a neutral monism. For this monism the ideas of sensations and of bodily things, of mentality and materiality, are but different ways of conceiving the basic elements, ways that are constituted by the elements being regarded in different contexts of functional dependencies. As Quine might have put it disapprovingly, ontology is resolved into ideology, in the process rendering many a hoary old philosophical problem, as Carnap may have put it admiringly, into pseudo-problems.

In a similarly direct fashion *EoI* is placed by the Preface to *Band 2* right away in the context of the lecture series Mach gave in Vienna from 1895 to 1901 which were predominantly directed to different aspects of the history of the physical sciences but also to that of psychology and to the logic of scientific research. Readers are thus familiarized without delay with an interdisciplinary program of research that has points of contact, on the one hand, with the conventionalist approach to scientific theory-building of the French philosopher and historian of physics Pierre Duhem and, on the other, to the biologically oriented evolutionism of the Viennese philosopher and educationalist Wilhelm Jerusalem.

The Preface of *Band 3* introduces *MiiE* by way of citing Einstein's appreciation of Mach's "historical-critical" reflections on the progress of physics. This particular approach of Mach's was evident already in his earlier "On the History and Root of the Principle of the Conservation of Energy" and combines, as the editors show, a strong allegiance to empiricism with the agenda of anti-metaphysical enlightenment. Mach's empiricism is grounded in his pragmatic-evolutionary understanding of human cognitive functions which prompts, to begin with, the instinctive adaptation of thoughts to the facts of experience, which then is refined in the course of conscious reflection in scientific work by the adaptation of thoughts to each other. Anti-metaphysical enlightenment falls out quite naturally as an end product of these reflections.

The Preface of *Band 4*, the collection of shorter pieces which, incidentally, was first published in English by Paul Carus in America, fittingly features suggestive remarks on Mach's proximity to pragmatism—explicitly recognized by as such by

William James—and chronicles the collection's expansion across the five editions from 12 to 33 lectures to. Mach, it is rightly stressed, was master of the popular exposition of scientific subject matter without condescending moralizing or simplification.

Further volumes of the *Studienausgabe* will be the *Die Prinzipien der Wärmelehre*, *Die Prinzipien der Optik* and three volumes of shorter writings on physics, psychology and physiology, and history and philosophy of science. The great majority of Mach's works will thus be available again and, if the present volumes are a reliable guide, more accessibly so than ever. In sum then, the new bottles render the old wine more quaffable than ever—which is just as well another Mach centenary is nearly upon us!

Thomas Uebel
(University of Manchester)

Index

F. Stadler (ed.), *Integrated History and Philosophy of Science*, Vienna Circle Institute Yearbook 20, DOI 10.1007/978-3-319-53258-5

Printed in the United States
By Bookmasters